your own longing Florry — I want/not to make. I want your arms — your lips & dear! I want you my & my dearest & & I ever your very own Paquita

*this
everlasting silence*

this everlasting silence

the love letters of
Paquita Delprat and Douglas Mawson
1911—1914

edited by
Nancy Robinson Flannery

MELBOURNE UNIVERSITY PRESS

MELBOURNE UNIVERSITY PRESS
PO Box 278, Carlton South, Victoria 3053, Australia
info@mup.unimelb.edu.au
www.mup.com.au

First published 2000

Letters of Paquita Delprat © Estate of Francisca Adriana Mawson 2000
Letters of Douglas Mawson © Estate of Douglas Mawson 2000
Introduction and annotations © Nancy Robinson Flannery 2000
Design and typography © Melbourne University Press 2000

This book is copyright. Apart from any use permitted under the
Copyright Act 1968 and subsequent amendments, no part may be
reproduced, stored in a retrieval system or transmitted by
any means or process whatsoever without the prior
written permission of the publisher.

Designed and typeset by Sandra Nobes in Weiss and Humanist-Regular
Printed in Australia by Australian Print Group

National Library of Australia Cataloguing-in-Publication entry

This everlasting silence: the love letters of
Paquita Delprat and Douglas Mawson, 1911–1914.
Bibliography.
Includes index.
ISBN 0 522 84870 2.

1. Mawson, Paquita, 1891—Correspondence. 2. Mawson, Douglas, Sir,
1882–1958—Correspondence. 3. Love-letters—Australia—Women authors.
4. Explorers—Australia—Correspondence. I. Flannery, Nancy, 1929– .
II. Title: Love letters of Paquita Delprat and Douglas Mawson.

808.8693543.

Foreword

IN 1997 I WAS the leader of the first Australian expedition to walk *unsupported* across the Antarctic continent to the South Geographic Pole. We walked fourteen hundred kilometers from the edge of the permanent ice shelf at Berkner Island to the South Pole.

Upon my return to Australia, and at a function where my sledge was handed to the University of Adelaide's Waite Institute for display alongside Sir Douglas Mawson's famous half sledge, I noticed Jessica McEwin, Sir Douglas's younger daughter, studying me intently while I was addressing the audience. She seemed to have a tear in her eye. Afterwards she told me quietly that I reminded her of her father: the same focus, commitment, intensity, idealism, struggle, and love for family and country.

I can certainly identify with his dream and his struggle to achieve it. The long years of uncertainty, organisation and fund-raising, and the even longer years of repaying expedition debt, were the same for me as for him. So too was the feeling of isolation from family and loved ones.

Too often in polar literature the explorer is pictured as facing the wilds without fear. Loneliness is never mentioned, nor are the feelings of the loved ones left behind who suffered just as much. Nor is the ever-present questioning as

Foreword

to whether you are doing the right thing or just being selfish. In all these things I can identify with Sir Douglas: the months of no contact, of wondering what was going on at home, and the questioning of one's reasons for proceeding with the dream in the first place.

The silence is deafening!

The only way to cope with the grind of the trail is with the support of those you love the most. I knew this and so did Sir Douglas. Unfortunately, this was never talked about in his day. This book, therefore, is a critical piece in the Mawson puzzle because it shows him suffering the same frailties and doubts as the rest of us. He was human after all.

With this in mind, the account in this book of how he and Paquita surmounted the isolation of their long separation nearly ninety years ago is outstanding. Followers of Antarctic history will be grateful to Nancy Robinson Flannery for locating these love letters and skilfully weaving together the complex story of those years.

Just weeks away from our next expedition, to traverse three thousand kilometers of the Antarctic Continent from Berkner Island in the Weddell Sea to McMurdo Sound in the Ross Sea—again unsupported (that is, using only our bodies to pull our sledges, without sails or dogs or motors)—my family and I renew our attempt to deal with 'this everlasting silence'.

PETER TRESEDER, OAM
Sydney, September 1999

For Ian, my love and my helpmate;

for the descendants of Paquita and Douglas, for their generosity;

and for all true lovers, that they may gain strength to weather the

doubts of any everlasting silences which come their way.

Contents

Foreword by Peter Treseder   v
Preface   xiii
Acknowledgements   xvii
Editorial Note   xix

Introduction   1
THE LOVE LETTERS
December 1911 to November 1912   19
November 1912 to November 1913   49
November 1913 to February 1914   119
Epilogue   131

Sources   145
Biographical Index   149
General Index   152

Illustrations

Unless stated, photographs are by courtesy
of family members.

between pages 12 *and* 13

Douglas and brother Will as children
Paquita as a child
The young Delprat sisters
The *Aurora* in Antarctic waters
    Mawson collection, University of Adelaide
Paquita's photo from above Douglas's bunk
Douglas stands on *Aurora*'s rail
    Adelaide Advertiser, 27 February 1914
A pre-wedding portrait
    Melbourne Punch, 12 March 1914
The wedding group at Linden
Professor Sir Douglas Mawson
Lady Mawson
Douglas, Paquita and baby Patricia
Paquita's parents' golden wedding anniversary

Endpapers: facsimile letters (extracts)
from Paquita, 21 April 1913,
and from Douglas, 30 October 1913

Preface

THIS IS THE love story of two exceptional young Australians. Paquita Delprat and Douglas Mawson were engaged to be married when Mawson set out as leader of the Australasian Antarctic Expedition of 1911–14. The letters in this book were written by the young lovers during their twenty-seven month separation.

Both correspondents later wrote books. In addition, Douglas wrote and co-wrote more than a hundred published scientific papers, sent many thousands of letters to correspondents around the world, and edited the scientific reports of Antarctic discoveries and observations. For her part, Paquita wrote articles for journals and magazines, gave talks to women's groups and on radio, and created short pieces of fiction for both adults and children. She, too, was a prolific letter-writer all her life.

Thus there is no shortage of primary resource material by and about these English-born Australians whose lives spanned the reigns of five monarchs, from Queen Victoria to Elizabeth II. Their combined archival collections reflect the significant scientific and sociological changes between 1882 and 1974.

I first read Paquita's love letters in 1991. They are held among Douglas Mawson's personal papers at the Waite

Preface

Campus of the University of Adelaide. Despite diligent searching, it was another six years before I located those written by Douglas during the same period. They were among another vast collection held by a family member—their presence unknown to her because the papers, letters and photographs were largely unsorted.

All the letters are handwritten, were in no apparent order, and some bear no indication even of year. All but two of Douglas's are on the printed letterhead of the Australasian Antarctic Expedition 1911 (some also carry the printed S.Y. *Aurora* tag), for Mawson continued Ernest Shackleton's custom of taking copious supplies of formal stationery to the remoteness of Antarctica.

The letters are published here, unabridged, for the first time. In the case of Paquita's fluent and sometimes anguished letters to Douglas, I believe only one to have been published previously.[1] Paquita used excerpts from some of Douglas's letters in her 1962 biography of him, published four years after his death, but few of the more personal passages were included. There is evidence in other correspondence that Paquita had wanted to use the more intimate passages to illustrate the tender and caring side of Douglas's personality, seldom seen beyond the family circle. That those passages were deleted appears to have been due to the combined forces of contemporary editorial policy and the sensitivity of close family members. It is much to the credit of the family that such reservations have now been overcome, at a time when the tone of biography is more personal and intimate.

My introduction and annotations to these love letters are designed to guide readers through the complex settings,

Preface

time factors and dynamics of the protracted courtship. While much has been written about Douglas Mawson and Antarctica, readers of these letters might find particular interest in two books written by the lovers themselves, Douglas's *Home of the Blizzard* and Paquita's *Mawson of the Antarctic*, together with a little-known book for children written by their granddaughter Paquita.[2]

The letters are presented in chronological order. I believe this to be a complete set, from the eve of Douglas's departure from Hobart on 2 December 1911 until his return from Antarctica on 26 February 1914. The only exception is the farewell letter from Paquita to Douglas, which he acknowledges on 1 December 1911 but which has not been traced.

Their correspondence was not a conventional exchange of letters, where one correspondent replies to the latest letter from the other. The extreme isolation of Antarctica in that era made such an exchange impossible. Some sense of the emotional ramifications of such a separation can be gleaned from the fact that during the twenty-seven month separation, Paquita endured twenty-two months without a single letter from her fiancé. Douglas was a little more fortunate, the intervals between receiving mail from Paquita being fourteen months and ten months.

Partly due to the frustration of each in writing to a silent wall, and partly to the consequent crescendo of doubts each had about the love of the other, the letters increase in uncertainty as the separation lengthens.

Paquita eloquently described her feeling of isolation: 'But this everlasting silence is almost unbearable'. In reply, Douglas, who had not received her letter for three months,

Preface

agreed that indeed the everlasting silence had been 'unbearable'.[3]

[1] In *The Oxford Book of Australian Letters*, edited by Brenda Niall and John Thompson, Melbourne, 1998.
[2] Paquita Boston *Home and Away with Douglas Mawson*.
[3] See Paquita's letter of 21 September 1913 (page 102) and Douglas's reply (page 124).

Acknowledgements

THIS BOOK WOULD not have been possible without the generosity of the Estate of Francisca Adriana Mawson (née Delprat) and the Trustees of the Estate of Sir Douglas Mawson in granting permission for the letters to be published. Their patient and sustained interest in and help with my project has been both commendable and appreciated. The use of Paquita Delprat's letters is also by courtesy of the University of Adelaide, the custodians of the letters.

My thanks also go to:
the Delprat/Mawson descendants for the use of family photographs and the loan of scrapbooks and correspondence; and Lennard Bickell, the late Sir Russell Madigan, Jenya Osborne and Ingrid Davis for permissions to quote;

Alun, Gareth and Pamela Thomas; Paquita Boston; Andrew, Sue and Jessica McEwin; Stella Vitzhum, Stroma Buttrose, Philip Ayres, Rob Linn and Richard Ferguson for reading drafts of the manuscript and for making suggestions;

Bill Scammell, Mark Pharaoh and Susan Woodburn of the University of Adelaide; Gerard Hayes of the La Trobe

Acknowledgements

Collection, State Library of Victoria; Roger André and Neil Thomas of the Mortlock Library, State Library of South Australia; Graeme Powell, Valerie Helson and Bill Tully of the National Library of Australia; Alana Birchall of BHP Archive, Melbourne; Margaret White of Egham-by-Runnymede Historical Society, England; Brian Anderson, Cathy Bennett, Barbara Brougham, Tom Hynes, Michael Kelly, Jo Toms and Ralph Wallace of Broken Hill—all for general assistance with research and access;

the many people who contributed written and oral memories for my forthcoming biography of Paquita, Lady Mawson —all of whom helped to build a background for these letters and their aftermath;

the Riverina Writers' Centre, Charles Sturt University, Wagga, for providing a retreat in their cottage where, beside the campus winery, the sap of creation began to flow.

To the commissioning editor of Melbourne University Press, Teresa Pitt, for encouraging my vision and enthusiasm, and to Jean Dunn, my skilful and ever-vigilant editor's editor; to the modern-day Antarctic sledging expeditioner Peter Treseder for sharing the family aspect of polar isolation and for writing the Foreword; and to my husband Ian for his never-failing help, encouragement and loving criticism, I also acknowledge my deep gratitude.

Nancy Robinson Flannery
Adelaide, September 1999

Editorial Note

THE FEW SPELLING errors in the letters have been preserved and punctuation mirrors the original, except that dashes sometimes used by Douglas to mark the end of sentences have been replaced by full stops. Some long passages, particularly in Paquita's letters, have been broken into shorter paragraphs. For the sake of clarity, names of ships, titles of books and journals, and words underlined for emphasis have been italicised.

While all Douglas's Antarctic letters to Paquita were written from the same place, perhaps best described as Cape Denison, Adelie Land, he used a variety of headings. Indeed he does not once use the term Cape Denison, although he later wrote in *The Home of the Blizzard* that 'the Main Base was finally settled at Cape Denison, Commonwealth Bay'.

Conversions

1 foot	0.30 metre
1 yard	0.91 metre
1 mile	1.61 kilometres
1 pound (lb)	0.45 kilogram

'this everlasting silence'

Douglas's period without letters:
late November 1911 to 9 February 1913 (14 months)
9 February 1913 to 13 December 1913 (10 months)

Paquita's period without a single letter:
April 1912 to 26 February 1914 (22 months)

They did not see one another from mid-November 1911 to
26 February 1914

Introduction

The First Meeting

HIS FACE BROKE into a broad, spontaneous grin as he turned. The smile was not for her, because they hadn't yet met; with his blue eyes twinkling, he beamed at someone close to her.

Paquita was enchanted by that warm, mischievous grin and the very tall, upright figure of the young man who flashed it, touching his hat to the recipient. She whispered to her friend Hester, 'Who's that?'

'Oh, 'Quita, don't you know? It's Douglas Mawson. You remember—he came back from the Antarctic a few weeks ago. Went with Shackleton. Quite a hero, apparently.'

Yes, she did now recall. Although the newspapers and her father and older sisters had enthused, nothing had prepared her for the magnetic appeal of the young explorer, nor in particular for that infectious grin. Chaperoned by their respective mothers, the two girls continued to circulate amongst the guests at the outdoor function. The 17-year-old Paquita Delprat was not to know just how much that smile, that tall figure, that boyish charm and sharp intellect would influence the rest of her life.

During the ensuing winter of 1909, Paquita kept to herself the attraction she felt to the 27-year-old scientist and

[1]

explorer, for she was but a child compared to her four older sisters. What chance was there that the illustrious Douglas Mawson—now being pursued by Adelaide's society matrons —would even notice her? Lica and Mary, about to graduate in medicine, perhaps? Or Elizabeth, the violinist making a name for herself in music circles; or Leinte, the next youngest sister . . . but not her, not young Paquita who had only recently completed her schooling at the progressive private college, Tormore House, in North Adelaide.

In August of that year, with her hair swept up to proclaim her new status of womanhood, Paquita went to stay with her father and Leinte at the BHP general manager's large home in Proprietary Square, Broken Hill. One evening, they were dinner guests at the Boyds' home on the opposite side of the square.[1] Also a guest was Douglas Mawson, in Broken Hill for a few days completing some mineralogical work[2] and giving an illustrated lecture at the Technical College on his Antarctic experience.

For the first time, Paquita and Douglas were formally introduced. Her flashing black eyes met his mischievous blue ones. She blushed; and he surprised himself by being fascinated. Always an easy conversationalist, with an innate ability to put people at their ease, Paquita chatted with Douglas about his university work and his Antarctic experiences.

The hostess, Mrs Boyd, observed the conversation with interest and pleasure. She had long held a soft spot for the youngest of the Delprat girls, and was heartened to see her coping so maturely as a vibrant dinner guest. While Douglas responded keenly to Paquita's charm, her father sat back to enjoy the interplay. His pride in Paquita mingled with a

Introduction

slight uneasiness at her sheer youthfulness, yet he reminded himself that he'd been enchanted by her mother when she was much the same age. Leinte, meanwhile, was amazed at the attention that Mawson was paying her youngest sister.

There seemed to be a mutual attraction . . . but Douglas and Paquita were not to see each other again for nearly a year.[3]

The Protagonists

Douglas Mawson was born in Yorkshire in 1882, emigrating to Australia two years later with his parents, Robert and Margaret, and older brother William. The younger Mawson's scholastic ability, together with a resourcefulness born of both nurture and necessity, enabled him to enter the University of Sydney at the age of sixteen and to graduate as Bachelor of Engineering three years later. With a growing interest in geology and an admiration for his mentor, Professor T. W. Edgeworth David (1858–1934), Douglas added a Bachelor of Science in 1904, before being appointed as lecturer in mineralogy and petrology at the University of Adelaide.

When Professor David was invited by Ernest Shackleton (1874–1922), early in 1907, to accompany him on his forthcoming British Antarctic Expedition, David recommended that Douglas Mawson be engaged as physicist. Writing of this period fifty years later, Douglas told of how, when Shackleton came to Adelaide for half a day en route to New Zealand to join *Nimrod*—the ship in which he would sail south—they

Introduction

met by arrangement on the steps of the Adelaide Club. In those elegant, conservative North Terrace premises, preliminary negotiations took place, and Mawson's initiation into polar participation was later confirmed by telegraph.

Douglas was away from Adelaide with Shackleton's expedition from late 1907 until early 1909. His achievements and displays of leadership engendered a passion to gain more scientific data from the south polar region, and early in December 1909 Douglas left for England and America to discuss the possibility of future Antarctic exploration. By the time he returned to Adelaide in June 1910, it had become obvious that, if he were to achieve his scientific goals, the best policy would be for Australia to mount its own polar expedition.

Paquita (Francisca Adriana) Delprat was also born in England, but not until 1891, nearly a decade later than Douglas. She was the sixth child and the youngest daughter of Dutch parents, temporarily living in London. Her mother-tongue was the formal Dutch of her parents. It was spiced with the vernacular Spanish of the family's servants and the village ribaldry of the Andalusian mining area to which her family had returned when Paquita was a toddler.

Her father, Guillaume Daniel (GDD), was descended from a line of distinguished Dutch courtiers, soldiers and clerics; further back was Basque blood, from which Paquita had inherited her dark eyes and black hair. From her mother, Henrietta, of a well-regarded French/Swiss/Dutch family, she had ivory skin. Paquita was born prematurely, a tiny baby who quickly outgrew the Spanish diminutive for Francisca, yet—except on some official documents—the

Introduction

name Paquita stuck for life. By the time Douglas noticed her, 'little Francisca' had grown to be a tall, elegant sapling, complementing his own wiry 'six feet three in his socks'.

Paquita's mining engineer father accepted an invitation from the fledgling Broken Hill Proprietary Company to come to Australia, and arrived in Adelaide in September 1898. The following year he was appointed general manager.[4] Henrietta and the younger children followed in RMS *Himalaya*, arriving in Adelaide on 8 January 1899. They spent their first night in Australia at the Largs Bay Hotel before proceeding to inland Broken Hill the following day.

Paquita's introduction to Australia was under conditions even harsher, hotter and more remote than those of the Spanish mining country. Even for the privileged family of the company's general manager, the lifestyle of Broken Hill was inescapably that of a nineteenth-century frontier town. The isolation and conditions engendered rapid civic development and forged strong bonds of loyalty and supportive good neighbourliness amongst its inhabitants. Building on her early-childhood experiences of rural Spain—where the Delprat family had learnt to shrug off the sometimes hostile reactions of the local people—Paquita learned in three years at the more friendly Broken Hill to be resourceful and compassionate. Although family letters reveal that Paquita could be quite mischievous as a little girl, and fun-loving as an adult, photographs of her in Spain and at Broken Hill show her as a particularly serious-faced child.

Together with some of her siblings (the older girls were at boarding school in Holland and did not rejoin the family until October 1900), Paquita attended the Convent School at Broken Hill, where she quickly mastered English. Yet for the remainder of her life she retained a slight—and to most people appealing—European accent.

Paquita's father bought a home in Adelaide in 1902, where her mother then spent most of her time, while the children attended school and university. As GDD travelled a great deal for BHP, he retained the use of the rambling general-manager's house at Broken Hill, but took every opportunity to be with the family at North Adelaide.

Delprat bought five acres of land at Brighton, a suburb of Adelaide, in 1909. The land had a frontage to the sea, and on the gulf-fringed sandhills a spacious summer house was built. He named it El Rincon after a mine and its house in Andalusia, where the family had holidayed while living in Cordoba.

The Romance Blossoms

It was therefore at Broken Hill, at Tynte Street in North Adelaide, and at Brighton that Douglas courted Paquita during the latter part of 1910. In the early days of the romance, young Willy (Dan) Delprat was flattered at the attentions of Mawson, until he realised he was being used as a conduit to his youngest sister. In later life, acknowledging this to be a ruse as old as love itself, Dan recalled the ploy with glee.

Douglas was infatuated, not only with Paquita but also with her large, happy, affluent family of stimulating, gregarious people. For too many years he'd lived a somewhat lonely and sometimes a spartan life, away from a proper

Introduction

home—a consequence both of his own ambition and of chance. Nevertheless, he now became aware of what he had been missing. The prospect of sisters who were artistic, intellectual, cosmopolitan and urbane, and a mother who was not constantly beset by hard work and financial problems, as his own beautiful and intelligent mother had been, fascinated him. Nearly three decades later, Paquita wrote of her siblings:

> All of us were imbued with a desire for new experiences, with their accompanying hardships. Exploration is a venture of the inquiring mind before it is translated into an effort and we were well prepared for it.[5]

One quiet, dark evening of December 1910, with the rhythmic sounds of waves lapping on the beach beyond the sandhills, Paquita and Douglas sat in what had become a favourite spot during some weeks of concentrated wooing on Douglas's part. On the wide verandah of El Rincon, Douglas asked Paquita to marry him. Without hesitation she accepted.

But the engagement needed the formalities of the era. In a long letter to Paquita's father, dated 7 December 1910, Douglas wrote:

> My Dear Mr Delprat
> and I hope with your consent
> *My Dear Father*
>
> It is a long story that I have to tell you and I beg your forbearance.

[7]

Introduction

> Love has run out to meet love with open arms—it is the ideal story. I hope you will approve for the persons referred to are your much beloved daughter Paquita and myselfall-unworthy [sic].
>
> Naturally you will be interested in my position and prospects for the future which, I must confess are not as good as I should desire when considering its reflection upon one so good as Paquita. There is, however, dogged grit in our family (though sorry I am to say it) which has dragged us over many obstacles and will never forsake me while I have my being.

Douglas then went on, over seven handwritten octavo pages of University of Adelaide letterhead, to outline his family's background in England and his father's unfortunate losses in business ventures in England, Australia and Papua New Guinea, and to describe how his mother had shown 'her grit by taking several boarders, of course in a very private way— Mother felt the tax, however, and it has permanently reduced her vitality'. He continued:

> Personally, I have never yet failed in anything I have undertaken and look forward to a bright future. The characters of my Father, Mother and Brother and myself will stand the closest scrutiny and nobody can say a word against us [on] any score. Our motto is 'honour first'.
>
> I have never loved . . . any women before nor spoken to any women in terms intending wedlock. I have drifted into an intense love for Paquita and if a life long union is not realised my life shall be for ever blighted, but this would not matter if you think it for Paquita's good. I must plead guilty that I have

Introduction

such poor means for supporting so precious a treasure but I believe this will be rapidly improved.

My salary at the University is but £400 per year and I make an uncertain amount by reporting on mines etc. I have only about £700 in the bank. Shackleton owes me £300—and the Hungarian Gold Mining Coy which should be floated now owes me £2,500 fully paid up £1 shares in a £200,000 Coy. I may make something out of Radium. I can look forward to any good Geological position in the Commonwealth with a fair degree of safety—also a chair of Geology at this University.

At the present moment I am looking forward only to your consent to my union sometime in the future with your treasured daughter Paquita. In all probability I shall go to the Antarctic for 15 months from the end of next year, then I shall return with your consent to my dear love.

I'll say 'Here's bushels of gold, love' and I'll kiss my girl on the lips;
'It's yours to have and to hold, love.' It's the proud proud boy I'll be,
When I go back to the old love that's waited so long for me.

Be merciful—

Yours very truly
Douglas Mawson

P.S. Mrs Delprat will be interested to read this.   D.M.

From Port Pirie on 8 December, Mr Delprat replied:

My dear Mawson

I received your letter this evening and I can tell you that you gave me something to think about. The result is that if I don't wish to miss this mail I have only half an hour left in

Introduction

which to write. I must be shorter therefore than I would like to be.

To come to the point. I fully approve of you as a son-in-law. Let me tell you that I don't know a better man than yourself in Adelaide to whom I would trust the future happiness of Paquita with greater confidence. It was very good of you to give me such a full account of yourself—and your people—and I am going to be equally open with you. If I am too abrupt, forgive me—the subject is of too great importance to use language which could possibly be misunderstood.

You know what a united family we are, and the only desire my wife and I have is to see our children happy. I really believe you will make Paquita as happy as she can be. That you are not wealthy does not count at all. If I tell you that when I married I had only £200 salary and nothing in the bank, you will understand that this does not concern me much. As long as you have grit and intelligence and the firm dogged determination to succeed in making a happy home for your wife, you are bound to succeed. I believe in you—and what is more, my wife believes in you, and she is a pretty good judge of character.

The only shadow on the picture is your Arctic Antarctic expedition! Do you think it is a wise thing to run these risks and expose your wife to the terrible anxiety and all its consequences when you go away for 15 months on a dangerous trip where you may meet with accidents or where your health may be permanently injured through hardships and exposure.

Do you think it is a fair thing to make a woman go through?

Introduction

Do you think it helps you to build up the home you want to provide her with? You have made a great name for yourself there already—what good can a second trip do you? While you were away, we would all be having a terribly anxious time —and it would be a terrible strain on Paquita which would not do her any good. Have you thought this well over? Would you not be able to find as wide a field for your energy in the inhabited world, with more profit to yourself and your's, and no anxiety to 'the girl you left behind you'?

Don't think that I want to make any conditions. Every man must fight his battle in his own way, but as your success is my girl's happiness—and if anything happened to you her life would be blighted—I am sure you will excuse me for putting all this bluntly before you.

Think it over again! I fully understand that to give up such an expedition, must mean a tremendous lot to you. Do you think you can do it? You want a great deal—it is worth the sacrifice? I am sure it is. Well if I had you here comfortably settled in an easy chair, we could talk the matter over comfortably and either one or the other would give in—but on paper it seems so matter of fact. Have a talk with my wife about it—and when I come to Adelaide next week we will have a final discussion.

It is terribly hot here today, and the flies are keen and active. The smoke from the works is just on my house—and this and a lot more makes me wish that I were in Adelaide just now, but I cannot possibly be there before Wednesday. I hope you will be there then. Let me know!

Yours very sincerely
G. D. Delprat

Douglas somehow circumvented Delprat's perfectly reasonable and, as it transpired, prophetic reservations regarding the planned Antarctic venture, and both the formal engagement *and* the Antarctic plans went ahead. Perhaps it was the disarming Mawson smile and charm which did it— man to man, or man to woman, or both.

Australia's press, not to mention society's tongues, devoted much attention to the engagement. The *Bulletin*, for instance, enthused that 'Douglas Mawson, the brilliant young scientist who covered himself with glory in the Shackleton expedition . . . has undertaken to lead one expedition to the South Pole and another to the church'.[6] Several papers confused Paquita with her older sisters, describing her as a medical graduate of Adelaide University, whereas she was less than two years out of school.

Meanwhile, Douglas, in writing to Sir Samuel Way about the engagement 'to the youngest Miss Delprat' enthused that 'Paquita is only 19, and is just one of the dearest and noblest-hearted girls you can imagine'. And Douglas's mother —who had recently suffered a slight stroke—wrote to Paquita from Sydney: 'I am glad to know that Douglas has found a girl he can love, as I know he is difficult to please . . . I love the girls as I have not a daughter of my own'.

The Partings

The first unalloyed excitement of betrothed togetherness lasted for only a few weeks, for in January 1911 Douglas left Adelaide for Melbourne then Sydney, before sailing to

*Douglas (right) with his brother Will in about 1884,
before sailing for Australia*

Paquita in about 1897, in Spain

*The Delprat sisters in about 1893, in Spain:
(from left) Lica, Liesbeth, Mary, Leinte, Paquita*

SY Aurora viewed from an Antarctic ice cavern in 1913, photograph by Frank Hurley

A 'grown up' Paquita, Holland 1912, framed by Douglas and hung above his bunk in Antarctica, 1913

*'Where's Paquita?' Nearing Adelaide, Douglas prepares to jump from
the* Aurora *to a pilot vessel, with Cecil Madigan alongside,
26 February 1914; from a cutting in Paquita's scrapbook*

A pre-wedding portrait of Paquita and Douglas in Melbourne

Paquita leaves Linden on her wedding day, with her father and dressmaker behind, 31 March 1914

Sir Douglas Mawson in London in 1914

Lady Mawson in London in 1914

Douglas, Paquita and baby Patricia in 1916

The family gathers in Melbourne in 1929 for the golden wedding anniversary of Paquita's parents: (from left, back) Theo, Fer MacDonald, Mary, Peter van Buttingha Wichers, Peter Teppema, Douglas, Milo Sprod; (front) Carmen, Leinte, GDD, Henrietta, Paquita, Lica. Willy (Dan) is absent.

Introduction

England again—all for further planning of and fund-raising for the Antarctic expedition.

'I am already heartsick to be away from Brighton', he wrote from Melbourne. In Sydney he presented a paper on Antarctica to the Australasian Association for the Advancement of Science; while in England and on the Continent he had a heavy and often depressing schedule of lobbying for funds, engaging expeditioners, purchasing stores and equipment, and selecting a ship and a monoplane.

It was as well that Paquita had been reared in a home where her father (one of the highest-salaried men in Australia at the time) led a peripatetic life, commuting by overnight trains and ships in schedules which to a less fit man would have been impossible: Broken Hill–Port Pirie–Whyalla–Adelaide–Melbourne–Port Kembla–Sydney–Newcastle–Melbourne–Adelaide–Port Pirie–Broken Hill.[7]

Douglas returned to Adelaide in RMS *Morea* on 22 July 1911. A joyful reunion with the patient Paquita was short-lived, however, because six days later her fiancé was off again by train to Melbourne and Sydney for press interviews, further planning with his committee, and discussions with parliamentarians and industrialists—forever chasing funds for the forthcoming expedition. During this time, he also chose more expeditioners from the many applicants clamouring to share the polar experience.

He was back in Adelaide for a few days in early August, before re-tracking to Melbourne by the overnight express. Then several days of precious reunion with Paquita in Adelaide preceded sailing to Perth in mid-September, followed by a visit to Brisbane early in October.

Introduction

Although Paquita found these leave-takings frustrating, she believed it was preparing her for the longer separation of the Australasian Antarctic Expedition (AAE), scheduled to run for about fifteen months from December 1911 to February or March 1913. At least during these preliminary absences, she was able to keep in touch by letter, telegraph and an occasional call on the newly connected Delprat telephone (which Douglas disliked using).

When Douglas told her about the acquisition of some pioneering 'wireless' equipment, which he felt confident would enable regular contact with Australia, Paquita felt reassured. The press, with equal optimism, was predicting 'daily communication from the Antarctic' and extolling the AAE wireless plans as 'a feature entirely new to polar exploration'.[8] Despite Douglas's impending isolation, it appeared that their love bonds would be well nurtured by constant communication.

Paquita continued to be cheered and supported by her close-knit if sometimes geographically separated family. She busied herself with learning more domestic skills; continuing her singing and piano lessons at the Conservatorium of Music; keeping her first scrapbook on Douglas's movements as reported in the Australian and overseas press; and skilfully hand-sewing many brightly hued calico bags—in line with AAE colour-coding for the myriad contents of five thousand bulk packages for use in Antarctica. These bags sewn by Paquita and known affectionately to the expeditioners as 'Paquita bags' were tokens of warmth, colour and civilisation in the frozen white world of isolation. To Douglas, of course, the Paquita bags were of special significance—talismen of

Introduction

her love for him. One red bag, in particular, was later to play a vital role in his survival.

The early summer of late 1911 heralded the longest farewell of all. The *Aurora*, a 165 foot steam yacht formerly of the Newfoundland sealing fleet, had arrived in Hobart from London under the command of Captain John King Davis. The thirty-one expeditioners, the stores and the gear were gathering, and the time came for Douglas to leave for Tasmania, after a Royal Society farewell dinner in Sydney.

Although reluctant to leave Paquita, Douglas was now keen to initiate the action part of the expedition. He was tired of all the detailed organisation, lobbying and fundraising. Perhaps this impatience prompted his final edict to Paquita and her family: 'Don't write much to me. When Captain Davis returns for us in the *Aurora*, I'll be far too busy to read many letters'.

Paquita tearfully said her private goodbyes to Douglas, but was persuaded not to attend Adelaide's public farewell functions 'as both my mother and Douglas felt that my emotions, not always under control, might be embarrassing'. Being a dutiful young woman of her era, Paquita bowed to the ruling, accepting that both her mother and her fiancé meant well. 'I have always regretted my absence', she wrote fifty years later.

Douglas was accorded a standing ovation at the Adelaide Town Hall. The cheers of all those people were gratifying, if a trifle embarrassing to such a modest man. But where was the person dearest to him?

It was a salutary lesson to both the lovers, an experience not to be repeated. On Douglas's next appearance at the

Introduction

Adelaide Town Hall, in 1914, Paquita appeared on the platform with him.

1. The Delprats had lived in this house when they first arrived in Broken Hill ten years earlier. Adam Boyd was underground manager for BHP.
2. Mawson's D.Sc. thesis (1909) was entitled Geological Investigations in the Broken Hill Area.
3. This scene is drawn from interviews with family members and my reading and interpretation of Paquita's two books and thousands of family letters.
4. See P. Mawson, *A Vision of Steel*.
5. *The Times of Ceylon, Sunday Illustrated* (Colombo), 25 December 1938.
6. *Bulletin*, 29 December 1910.
7. G. D. Delprat, Diaries (National Library of Australia, Canberra).
8. *Observer*, 28 January 1911; *Adelaide Register*, 21 October 1911.

The Love Letters

December 1911 to November 1912

THE S.Y. *AURORA* LEFT Hobart late in the afternoon of 2 December 1911, following more civic and vice-regal farewells, a special service in St David's Cathedral, and receipt of messages from King George V, and Alexandra, the Queen Mother.

To fill the long separation and to give Paquita a taste of European culture, Henrietta, despite a distaste for long sea voyages, planned to take Paquita and three of her siblings abroad early in 1912. Douglas knew of and applauded this plan, although Paquita's voyage to the northern hemisphere would take his love even further away from his own self-imposed exile in the extreme south.

Paquita continued to take press clippings from many papers and journals. While the Mawson expedition had no intention of joining the race to reach the South Pole, the nation's imagination and interest had been stirred by the scientific objectives and the sheer courage and drama of the undertaking. For such a young nation, only a decade after Federation, Douglas's plans were indeed ambitious. The country seemed ripe for a hero and was justifiably proud that Professor David, in the aftermath of the Shackleton expedition, had said that Australia had in Douglas Mawson a man of infinite resource, splendid physique and a real leader.

Paquita's emotions must have been mixed. Enormous pride mingled with concern, and the first inkling of just how great the loneliness ahead may be. At least during the first few months of the separation, she could expect to receive some letters from Douglas—despatched by mail steamer from Hobart during the final preparations, then by a merchant vessel returning from Macquarie Island, and finally by the *Aurora* returning from Antarctica after delivering her men and cargo.

Tasmanian Club, Hobart
Friday 1 December 1911

Sweetheart and More—Far More—

I have a great longing to say something to you but I cannot in a letter communicate my feelings. You may be sure that I am going away this time far happier than last when there was no gem of priceless worth awaiting my return. You may be sure I will look after myself compatible with a dutiful endeavour to accomplish.

I had to cease writing this afternoon when the Premier came in[1] and now at 12.45 am Saturday send a few more lines. Your farewell letter is to hand[2] and I love to feel an intensity of feeling in your utterances.

I am very sleepy and you must excuse. We have had a farewell tonight by the Premier—a great crowd of people attended. I gave 2 lectures here to raise cash this week [and] got £60 clear. Have just sacked Joyce.[3]

We have a fearful quantity of deck cargoe[4] and are leaving behind some of our stores. With good weather we

[20]

shall reach the Antarctic without fail—very heavy weather may wash some of our cargoe overboard. The men are all good.

Now just one big hug and au revoir until you get a billet doux from Macquarie Islands.

Douglas

[1] Sir Neil Elliot Lewis (1858–1935).
[2] This letter has not been traced.
[3] Ernest Joyce, a former seaman, had served with Scott and Shackleton. While it is known that Captain Davis disliked Joyce, the reason for Mawson's sacking of him is unclear.
[4] The *Aurora* was carrying close to 600 tons, 200 more than her registered tonnage.

Nearing Macquarie Islands
Sunday 10 December 1911

Angel,

I have been thinking such a lot of you these past few days—between the rolling and pitching of the ship—between watch and watch. How grand it would be to fly back to you, even for a few brief minutes. The discomfort of a boat of our class laden as we are and in the heavy weather through which we have passed, drives all my longings into one channel—peace and you.

Our journey has been lengthened by a gale and encumbered decks—we scarcely thought to get through Tuesday last, but providence has asserted itself right from the beginning. The Starboard bulwarks, motorboat and part of the bridge have been broken up—repairs are however well under way. It is calm to-day but a large swell running. Tomorrow morning we should be anchored at Macquarie Island.

Those telegrams from yourself and Lica and Leinte greatly appreciated—at the last moment I got a few small books at Hobart and am posting back from Macquarie Island. To *Mother* (yours) I am posting a butter knife with our mark upon it. It will do for a book cutter?

I have lost the letter from your father—it was very nice indeed. I love you *all* so much! Excuse me blurting this out once again. Give my warmest affection to *Lica*.

The *Aurora* rolls day and night without ceasing—I have been trying to sleep on a couch and almost nightly find myself on the floor a few times during my watch below. The seas breaking over the ship have deluged everything—it is a horribly repulsive feeling to be dropped from the couch into 2″ or 3″ of water flowing backwards and forwards on the floor. Have not yet had my clothes off and only one rinse of my face—no hair brushing. You can imagine I am looking rather shaggy. We have managed to keep the dogs alive so far.

From Lovelace, little transposed:

> *Tell me not, sweet, I am unkind*
> *That from the nunnery*
> *Of thy chaste breast and quiet mind*
> *To Antarctica I fly.*
>
> *Yet this inconstancy is such*
> *As you too shall adore,*

*I could not love thee, dear, so much
Loved I not honour more!*[1]

Do read the 'Patchwork Papers'. Don't you like Elizabeth Browning?

I must go on duty now my Love

    Douglas

---

[1] From Richard Lovelace (1618–58), 'To Lucasta, On Going To The Wars'.

The 120 ton steampacket *Toroa* left Tasmania five days later than the *Aurora*. *Toroa* had been chartered to go as far as Macquarie Island, a sub-Antarctic Australian dependency 850 miles south-east of Hobart. By the time the smaller vessel arrived with some of the expeditioners and the requirements for a land base to be set up there, Mawson's advance party had carried out preliminary reconnaissance.

While Paquita had been briefed in advance by Douglas (always the teacher) of the dangers of even reaching Antarctica, she had little idea of the full range of difficulties ahead. Douglas's first three letters after leaving Hobart give little indication of just how rugged the early part of the storm-ridden journey had been. Perhaps he was saving her distress in the same way that, within a few years, Australia's soldiers would do in their battlefield letters to mothers, wives and sweethearts.

Paquita did receive a belated indication of the potential hazards of the Southern Ocean. While in Holland, she pasted into her scrapbook a Sydney press report of April 1912 which read:

> An examination of the antarctic exploring ship *Aurora* in the dry dock shows the Mawson expedition narrowly escaped disaster when the vessel bumped on a submerged rock at Macquarie Island on December 11. The heavy keel had been torn away for about 15 feet. The whole of the bottom shows rough usage, and in one place it was smashed in, although the shell of oak is 15 inches thick in the spot.

Macquarie Island
15 December 1911

Darling,

The last few days have been very strenuous ones but I like it—I am in my element. Hard physical work agrees with me. I have only had one rinse of my face since leaving Hobart and there is very little skin on my hands now. Have brushed my hair twice. You would scarcely recognise your Dougelly. Everything is going well, though we have a large contract erecting this station on top of a high hill. Generally speaking all hands are working well. The general reports are in the newspapers and I need not repeat.

Had intended writing a short note to your sisters, Willy, Mother, Father etc. etc. but time now prevents. Take it from me we are going through with it well and I shall return to you in Autumn 1913. Letters from all the family received by *Toroa* much appreciated.

Believe me Paquita Your devoted Douglas

December 1911 to November 1912

The next two letters to Paquita were sent back on the *Aurora*, arriving in Hobart on 12 March. The letters were then forwarded to Paquita in Europe by Conrad Eitel, the expedition's secretary and press correspondent based in Tasmania.

Adelie Land[1]
3 January 1911 [1912]

Paquita, Just Paquita: that is far more than all the dearests, darlings etc to me now—Paquita stands for so much to me now above all times. I am up against it and my very policy has to take into consideration at every turn what would be the best for Paquita.

This morning I wished that I had never spoken to you of my love—that was because a large proportion of failure appeared to stare us in the face (we had had the worst possible luck for some days past)—now that we are commencing to succeed I am just glad to write to you and remind you of my devotion. Things looked so bad last night that I could do nothing but just roll over and over on the settee on which I have been sleeping and wish that I could fall into oblivion without affecting you, darling.

This morning a great turn in events took place, and I now feel sure that we can complete the remainder of our programmes, though somewhat modified. Providence came forward at the eleventh hour and made a heaven for us.

At this moment I am in the engine room getting a warm up after a cold day's work in a keen wind—hence sundry oil spots on this paper. Before commencing I washed my hands in a pint of water removing all the loose dirt in case it would unduly smudge the sheet—this is the second wrinse I have had since leaving Hobart. Do not, dearest, think that I am so far a savage as to quickly forget the obligation of washing—my neglect in this respect has been due to our shortage of fresh water—we had intended watering at Macquarie Island but were unfortunate in getting only half the quantity intended before we found it expedient to leave.

As you will have heard [our] landing at Macquarie Is. was most successful. Fortune favoured us south until we reached the packs—then we had a run of bad fortune. We met heavy impenetrable pack in several directions and failed to break through to the land. Much of this disappointment and trouble I find today to be due to an undue reliance I had placed in the accounts of Commander Wilkes who made explorations here in 1840. His accounts are largely erroneous & misleading.[2] I had thought to lay a first base as far East as possible even as far as the 156° E. longt. After our experience I have now arranged to consolidate parties 1 and 3 and make a stronger base at Adelie Land, near which shores we now are. I have decided to rule Murphy out and to strengthen Wild's and my own parties. Murphy I will place in charge of my hut in my absence.

We expect to be landing the main party tomorrow or next day and everything I believe will go well. Already we have made important discoveries and, Oh my Dear, I am beginning to live again after a period of several days of impending evil and disaster. At 4 am this morning I had decided to turn back 50 miles and drive the *Aurora* into the pack never to come out unless successful in reaching land. I thanked God at 6 am when, the weather clearing, we discovered an unexpected glacier tongue and a lead to the south.

All on board are in the best of spirits and going strong. You will be in Europe ere this letter gets to you—the excitements will be very great but I hope Paquita *occasionally* thinks of Dougelly—wrap me up in your arms sometimes dear and warm me. Perhaps it is your love warmth that already protects me from the cold, for I doubt if I feel it so much this time as last.

I would dearly like to be with you to visit some of the cities of Europe—we must in 1913—I can almost fancy myself now with you in Paris & London—what fun it will be—or will you be bored with me? I sometimes think that I am much better out upon a lonely trail for nature and I get on very well together—I feel with nature and revel in the wilds. Here within a gunshot is the greatest glacier tongue yet known in the world—no human eyes have scanned it before ours. What an exultation is ours—the feeling is magical—young men whom you would scarce expect would be affected stand half clad without feeling the cold of the keen blizzard wind and literally dance from sheer exultation—can you not feel it too as I write—the quickening of the pulse, the awakening of the mind, the tension of every fibre—and this is joy.

In whatever pastures it lies may yours be joy also. Well dearest I do not expect anything will happen to detain me another year from your loving self and I expect to return with the *Aurora* in Autumn 1913. You will not hear of us till the end of April for I find the season is later here than I expected and our programme is likely to detain us until then.

I intend then to give lectures in Australia and then go home[3] to lecture and publish—I don't know of anything that will prevent this—I desire to have you with me on this trip and I hope that you may accompany me—to look after me as I think I have heard you say. However, I can't leave Australia for 2 or 3 months after returning there. If you do not care for travelling we can arrange otherwise. I live for you, so please arrange and order.

Time is so very circumscribed that I can't attempt to write to all the family so please pass on my best love to them all.

Douglas

---

[1] Adelie Land, part of Antarctica, was first sighted by the French navigator Dumont d'Urville in 1840, and named *Terre Adélie* after his wife.

[2] Charles Wilkes spent forty-two days in Antarctic waters in 1840 without once setting foot on land. Douglas later modified his comments. 'Considering that his work was carried out in the days of sailing-ships', he wrote in *The Home of the Blizzard* (1996, p. xxxii), 'with crews scurvy-ridden and discontented, it is wonderful how much was achieved'.

[3] Used in this context, 'home' would have meant England.

Adelie Land
19 January 1911 [1912]

My Paquita,

Just a hurried line to say goodbye for a year—we are just about to go ashore at winter quarters, having landed all the needful.

What has happened in the last fortnight you will hear from the press. Suffice it to say that everything has gone off well though we had hoped to find a more rocky coastline.

You will not get this letter until the end of April or May and by then we expect to have the wireless in operation so you may hear earlier.

We have made a successful landing and I don't anticipate anything in the nature of disaster. Your wandering Dougelly will return with the Olive Branch to his haven of rest in little over a year's time. Of course that is if he is still wanted—from what you have said anyway, he is coming back to enquire. You will be quite a woman of the world then—perhaps quite too fine for me? Eh. Well don't let that be, for in this stern country of biting facts ones love gets frozen in deeper and there is plenty of time here to think over all the happiness that may be ours. The very fact of your loving me seems all that I want and I could live always in that beatitude.

Know O'Darling that in this frozen South I can always wring happiness from my heart by thinking of your splendid self.

There is an ocean of love between us dear.

Your loving Douglas

Early in February 1912 Henrietta, Lica, Leinte, Paquita and Willy Delprat left for Europe in a German twin-screw steel steamer of nearly 8000 tons. The other two Delprat sisters, Mary and Liesbeth, had preceded them. One of the purposes of the trip was to put together Paquita's trousseau—part of a well-bred young woman's unofficial dowry. So well and so abundantly was Belgian household linen bought on this visit to Europe that some of it is still used by Paquita's descendants nearly ninety years later. In a letter to Douglas from Holland late in 1912, Henrietta Delprat wrote: 'We almost want a ship to ourselves with . . . all the wonderful things Paquita has collected'.

Although she'd visited Holland as a child in 1896 and 1898, this was Paquita's first opportunity as an adult to become better acquainted with the country and the culture of her forebears. Broken Hill, Brighton and North Adelaide must have seemed far away.

Dampier Seydlitz
Norddeutscher Lloyd, Bremen
28 February [1912]

My dear, dear Douglas,

It is no use, I cannot wait until October to write to you.' Here I have been longing & longing for you and now I must write. You said truly when you said that a sea voyage was redolent of love and longing. It is. I lean over the side & in the water see you—oh I think I love you now even much more than when we parted.

December 1911 to November 1912

How simply glorious it will be when we are on our trip together. Dougelly, think of it. Just you & I seeing all these sights I have seen—Colombo with its exquisite colouring, dirty old Port Said—where it rained nearly all the time—and now Messina with its pitiful ruins. We are just out of the Straits.

There is a young French couple on board whose doings interest me so much. He is a little like you from the back but otherwise not. Still they seem so very happy that I sometimes think the time will never come when we are so.

I have enjoyed this trip so very much. Our cabins are beautiful, the weather ditto—although Lica & Mother have managed to be fairly sick. We are waited on very well & the passengers on the whole are very nice. There are some who I can hardly speak to politely & others who are— rather interesting, Dougelly dear. One man I have particularly chummed up with. He is married but rather young & we get on very well. He misses his wife & I my Dougelly, so we get consolation! But whenever we are talking I have always the feeling I could easily help him overboard if it would bring me you just for a few minutes.

I am longing for news from Australia. At Fremantle I got a wire from Miss Tomkinson that a wireless [message] had come from Macquarie Island & you—Feb 2nd that was. Then in Port Said I found a newspaper telling me that several messages had come through from Macquarie Island. Still they all are vague & I am anxious to see my papers when they arrive. Mr Eitel said he would send me all the wirelesses that came.[2] Oh darling we *are* far apart aren't we? Does it ever come to you with a rush. But you have so much to do. When I start studying again it will be better. If only nothing is happening to you but I think I should feel it.

I have been speaking quite a lot of German with the various people here. They are mostly German & half of them wool-buyers. All have voracious (!) appetites and drink heaps of beer & cup for all meals. I have never seen anything like it.

Tomorrow we arrive at Naples. Not long enough for a visit to Pompei. That will be for next time. Then we stay four days at Genoa to take the steamer for Gibraltar which calls at Algiers on the way. We are looking forward to that. It will be interesting. Our house is taken in the Hague & we only stay about 3 weeks in Spain.

I love this ship life & am a very good sailor. Even Leinte was sick in the Bight but I never felt it. Poor Lica had a terrible time. She could hardly walk at Fremantle.

Well my love. I have hanging over me your words that you would not have time to read many letters when the boat came back for you. But you will find time, I am sure. Anyway, I cannot help it. I love you & want you & writing to you is the next best thing. I have written before but always knowing I would tear the letter. This one I shall not.

It is no use wishing you success because when you get this you will be coming back and it will be nearly over. But I know you have had success. I know my Dougelly which says the same thing.

December 1911 to November 1912

With both my arms & my hearts love to you & in imagination my lips to you.

> I am
> Your Paquita

My thoughts you know are always with you.

[1] The *Aurora* was due to return to Antarctica in October, hence the first scheduled opportunity for mail to be despatched to Douglas. Paquita addressed her letters c/o Eitel in Hobart. In the event, the *Aurora* left Hobart on 26 December and Douglas did not receive the letter until 9 February 1913.

[2] 'Wirelesses' in this context meant messages sent by wireless telegraphy.

9 Wagenaarweg den Haag
Monday 1 April [1912]
in bed

My dearest Man

I read a most worrying paragraph in the Sydney paper to the effect that one Captain Hatch was taking a mail to Macquarie Island & also that there was one for you.[1] It may have been a misprint, although it was in two days following.

How terribly far I am from authoritative news of you. No news from you yet by wireless but will have your letter from *Aurora* this month. Oh, I hope you have written warmly in the midst of all that fascinating ice. I will be almost afraid to open it in case you wrote when

This Everlasting Silence

specially tired—but however you write I know what you feel.

I have already sealed up one letter for you. We didn't go to Spain! We change our plans often don't we? But one plan is fast and sure—we will be in Adelaide before you arrive—only now I don't know whether to expect you in Hobart or Fremantle. Fix up the wireless on the boat & let me know that way.

Well dear where was I? We got to the Hague safely. Stopped in Switzerland a few days en route & oh, it *was* cold there. I felt as if I was with you. Poor Lica does hate the cold but I don't.

We are now settled in our narrow, treble-detached, three-storey gardenless house. It is just big enough, with a room or two for visitors. Lica & I have just returned from a weekend to Amsterdam to hear a beautiful oratorio & stay with relatives. Such a real Dutch house. You shall see it when we come here together. I wonder if that will be next year or if you will come to Europe alone. I am impatient to know the future.

I am sitting comfortably in bed. East west, home is best! I shall like to go away later on to come back to you again & be home. Will you also tuck me up warmly like my mammy does & kiss me goodnight? Then you can get into your nest and I'll get out to tuck you up and so on until we both fall asleep on the floor!

This isn't telling you about Holland. I always return to you.

Well, we *love* it. So beautiful in its colouring. Everything is gradually turning into green. The tree stems were green and the boughs brown, now that is reversed and the grass also is greening. On the way to Amsterdam we passed through field after field of bulbs of all colours and sorts. The tulips are not all out yet, mostly hyacinths and daffodils. I wonder how they get all those beautiful colours. Purple blue. All shades of red and all in such neat rows. Each farmers paddock so to speak separated from his neighbours not by a fence but by a canal of rippling water generally edged with bare pruned willows. Dougelly dear, you must see my Holland.

Well, I wonder will you benefit by my sending this by this mail. I dont see what boat would venture to you. Still best be on the safe side. I'm thinking of you hard every night & wondering whether you are hungry or cold or lonely. If you were only here now. How your Paquita would warm you. I don't feel cold here at all. Must be my warm heart.

Mother is an angel & whenever I feel worried about your wellbeing tells me earnestly she would feel it if anything was wrong with you. We must be very happy together to make up for this year.

I am learning lots of things. We have table-linen patterns here! Would you like a tablecloth with a border of ducky chickens? When I think of getting things like that together I feel happy but when I think of my other half so far from me I lose all pleasure in them. My love, my love, how I miss you. I close my eyes and lift up my lips but feel nothing. How very far you are. I know you are liking your work & doing it well. You do most things well. Especially when I'm in your arms in the study at Tynte

Street you do things well. I wish I were there now, my dearie for always.

    I am your Paquita

[1] See Douglas's letter of 22 June 1913 (page 81).

9 Wagenaarweg
Haag
10 May [1912]

My Douglas, mine.

I'm feeling so absolutely healthy and happy tonight despite the distance between us that I want to write to you. I wrote you a letter on your birthday but it wasn't a success. I know you thought of us thinking of you then. And we did. It was you all day long. Lots of aunts and things sent flowers. There was one huge bouquet just like a wedding one! Then one aunt gave 'Us' a silver old Dutch spoon. They make a lot of fuss over a birthday here more than English do.

I hope you don't mind but I'm quite Dutch. I've never been so patriotic as now. I'm so awfully proud of my country and next year when *we* are together *we* must find time to come here, so that you can meet all our relations (& there are a heap) & see my Holland.

I got your letters safely and just when I was wanting something from you.[1] How different our lives are at present! My man, I wish I had been there to help you when you were so worried before you landed. What a lot we shall have to tell each other when we meet again. I want to come with you next year to Europe. I don't think anything will prevent it. You can't think how glad I am we came away for this year. We seem to be in quite a different world altogether. I don't think I could ever live happily in Adelaide again as before—that is if it wasn't for you, of course. When we have a home of our own it will be quite different. I'm picking up such a lot of ideas! I've got a book for them. And houselinen is busy being picked! I've got some old Dutch brass already and have my eye on more. Our house will be *the* house in Adelaide, in Australia. Douglas dear, aren't you glad we've got each other? I'm feeling the fellow feeling so very strongly tonight. No more boarding-houses for my man. No more nasty dinners and having to go away every night. Always together—I'm a very jolly person to have about!

I hope everything goes well with you now. We aren't worrying about not hearing per wireless yet. But I hope we do soon hear. Scott will return about the same time as you. I hope you come first. He will be disappointed at Amundsens getting there [to the Pole] first. How thankful we are that you aren't bound for there. Don't you go and stay away another year! You're under contract to return next year less than a year from now.

Mother and Leinte have gone away near Amsterdam for tonight and Willy and I are quite alone. Last week Leintie and I were in Amsterdam to witness a gala performance at the theatre in honour of our Queens visit there.[2] She generally lives here [in the Hague]. We saw her first in the street. She has a sweet face but not beautiful. They say her manner has changed a lot since the child came.[3] At night with all her diamonds & lovely dress etc she looked lovely. My uncle with whom I stayed is in the ministry & had places in a box where we saw everything beautifully. You will know the like.

Lica is following a course in Dresden. She is to be away a month but perhaps goes on to Berlin. Lizabeth has been & gone.[4] Mary comes the day after tomorrow. Its horrid without Lica & lovely to have Mary again. Lica is a duck. When we are settled we shall ask her to come and stay with us for a few weeks! I'm reading French with one aunt and German also. Feel quite at home in the latter.

Willy is growing up. He fell in love on the boat and has been repeating the experiment ever since. Unfortunately it makes him sometimes grumpy. Leinte is also a duck. And as for Mother I don't know how I shall ever leave her even for you. Yes I do though. There is only one you. I'm sure you don't love me as I do you. Women always love the most and miss the most. Well I wouldn't like you to miss me as much as I do you.

With my whole heart & my lips your Paquita

[1] Four letters (10 December to 19 January) were carried on the *Aurora* on her return from Antarctica, and from Australia to Holland by passenger steamer.
[2] Queen Wilhelmina of the Netherlands (1880–1962).
[3] Possibly refers to the birth in 1909 of Princess Louisa Emma, who reigned as Queen Juliana from 1948 to 1980.
[4] Three of Paquita's four older sisters.

The Hague
14 October [1912]

My Douglas,

First of all I love you even more than when you left & there has not been a day—an hour almost—that you have not been in my thoughts. You will have a warm welcome on your return—my arms are open for you already as I think of it. Not only from your me but from our Mother and the others. I'm quite spoilt! They have all tried to make up for your absence of course in vain but still it was & is angelic of them. But I had no idea I should miss you like this. Dearie, don't go away again. I'm longing to hear you say you've wanted me often.

I send you several *Illustrated Londons*. The contents may interest and it will give you an idea of whats been happening to the world since it was made dull by your departure. You needn't bring them back! Don't laugh at the contents of my little box to you! I did so want to send you something and knew not what. So eat the sweets and don't hurt your teeth & think that I have been eating them off & on all these months (!) I got the cuff-links for you in Colombo. You *are* a hard person to send things to & the circumstances are worse. I thought & thought ever since you left what I could send. Do you recognise my own mascot which I've sent you? I've slept under its [wings] for over five years & now it goes to my large & dearer mascot. That you must bring back. Do you like the photos? Snaps I have not many as the photographers of the family are continually absent.

This Everlasting Silence

And now for news. My own, what a heap I shall have to tell you & you me. Don't I long for the first quiet hours together. I make no promises to meet you as I can only answer for myself & not my family on whom it depends. But my heart will be there if not in flesh in spirit.

Lica & Mary will not return with us. They are for a year with Professor Wertheim in Vienna & like it immensely. Mother and I have been to visit them & love Vienna with its mixture of old and new civilization & its almost savage peasants. The people from the Tirol do look fierce sometimes. Mary failed her first F.R.C.S. exam but has wisely given it up & is doing practical work in gynaecology. She is going to specialise in that. Lica is just as ducky as ever. Leintie left us this morning for a week or two in Paris with an aunt. She goes to Berlin to Liesbeth in Nov & London the month after. Willy is at school and very much between a man & a boy. You should have seen him on the boat! Liesbeth (I tell you of all your sisters) has been here twice & now is in Berlin again. She is absolutely sick of being alone & returns [to Australia] with us. I did not know her at all for what she is. She left when I was still at school. She is one of the best. Mammy is absolutely just as sweet & really Mothery as usual. How very very fortunate I am to have such a family. Theo has another daughter & may come over next year for a certain event if it takes place then.[1]

Your loving me has seen a good deal, heard more & has learnt a lot. She is wiser in many things & loves you *much* more than ever. I shall have heaps to tell my man that I don't know how to write. I've seen Milan Lugano

December 1911 to November 1912

Lucerne Wisbaden Vienna Hague Amsterdam etc etc & we purpose travelling to Naples & staying in Rome for a week or two before joining the *Roon* Feb 26th for Australia and *you*. Have left Paris and London to your guidance!

What else can I say? Don't for one minute think that sightseeing has been all I've done. I'm taking singing & French lessons & am reading a lot & sewing & all sorts. I've realized much more how very little I know but I've found out a little now to set about to learn more.

We shall be very happy when you return with the separation behind us. Dear, I know you have done good work down there in the cold—you & your little hutfull of men. Things must have been hard now & then & perhaps you have not been able to do all you wish but remember that our aims must be the higher because we can never reach them, & you have done your best. We are all proud of you. Don't be disappointed if you haven't done all you wanted to. How often we here have pictured you in your hut and sledging. I'm glad you don't feel the cold as last trip.[2] Of course it is my love that does it. I warm you every night. You are safer there in a way than many here. You will see in those papers a terrible wreck—mining disasters & now terrible cruel war.[3]

Oh how I hope it has done you nothing but good! You promised to come back fatter & better. It's no use telling you to rest on the *Aurora* on its return voyage. Your book & reports will fill your time. But when we are married then my turn will come and Paquita is going to look after, scold and cuddle her Dougelly just as much as she pleases & as is good for him. So come back resigned my dear, I shall never let you go again & you will have me for the rest of your days. Can you stand it? We are going to do heaps of good work in all sorts of ways.

I can't stop just yet. Is four letters too much? I have so wanted to write more but it would be all the same song. Of how we here have thought of you & how interested all my aunts & things are. And how many pretty things I have for *our* home, the first home in Australasia just as the head of it will be the first in the world to the other half of the head. Rather involved but perhaps you can untangle it.

Now I shall stop. You are coming back to the warmest & lovingest heart that ever beat for its other half.

I can almost feel your arms round me & involuntarily as I write lift my face to yours. Seventeen months without one caress! One embrace. We *shall* have something to make up for.

> With my hearts whole love to you my lover
> from your Paquita

[PS.] You will have later news than I can give from Campbelltown.[4] Am afraid Mrs Mawson is not so well. But hope for the best darling.

---

[1] Theo, the eldest of Paquita's siblings, was at this time working in South Africa with the De Beers Company as a metallurgist.

[2] The 1907–09 British Antarctic Expedition with Shackleton.

[3] The first Balkan War (1912–13), with Montenegro, Serbia, Bulgaria and Greece attacking Turkey.

[4] From Campbelltown, near Sydney, Douglas's parents and his brother Will followed the AAE exploits with interest but mis-

givings. Paquita and her future mother-in-law kept in contact by mail.

By this same post, Henrietta Delprat wrote to Douglas that she was glad that:

> ... the time has come to write again; not that you have ever been out of my thoughts, but it is a different thing to know that we will be in communication with each other again ... I am glad also for Paquita's sake, for although she has been quite herself—always bright and cheerful as usual—I know she has missed you very, very much ... Goodbye my dear boy ...

Lica, from Vienna, wrote to Douglas expressing disappointment that she wouldn't be in Australia when he returned, to see his and Paquita's happy faces: 'There is a big picture of you hanging in her room and we've got to say nice things about it and to assure Paquita that she is right when she says "there is only one Douglas in the world". Well, dear brother, I mustn't take up any more of your valuable time'. Mary also wrote from Vienna at the same time and in similar vein, signing herself 'your loving sister'.

Meanwhile, during the southern winter and spring of 1912, Douglas wrote no letters to Paquita from Adelie Land, knowing that there was no way of sending them before the *Aurora* returned early in 1913—but his extensive diaries and logs, and those of his men, detail the story.[1]

The wireless communication had been a bitter disappointment; the weather conditions were even worse than

This Everlasting Silence

could ever have been imagined; within the sturdy hut they had built, the men kept busy with preparations for the early summer sledging journeys; and routine scientific observations continued, despite frequent blizzards with winds up to and sometimes exceeding 100 miles an hour, and air temperatures as much as 60 degrees Fahrenheit below freezing point.

Although Douglas and his companions looked forward to the increased exploration activity after their enforced hibernation, they also counted the days to the return of the trusty little ship. And none more so than their leader, eager to return to Paquita.

Morse code messages were sometimes reaching Australia from the operators of the wireless station on Macquarie Island ('the most southerly wireless station in the world' proclaimed the press). Not until the end of September 1912, however, could Walter Hannam—the main wireless operator at the Adelie Base—make contact with Macquarie Island, and even then could receive nothing at all in return. Night after night Douglas's messages ended with 'we have caught no signals yet'.

At least the outside world, via Conrad Eitel's press reports from Hobart, was now hearing that personnel at the Main Base, Australasian Antarctic Expedition, were safe and were fine-tuning their plans for the summer sledging journeys. In Europe these reports were clipped and added to Paquita's scrapbook.

Because Frank Wild's party, left by the *Aurora* more than a thousand miles west of Commonwealth Bay, had no wireless equipment, nothing could be heard from them. And

December 1911 to November 1912

as the Main Base had been unable to pick up any incoming messages from Hobart via Macquarie Island, Douglas did not even know where Captain Davis had put Wild ashore before sailing back to Hobart to escape the ice as autumn set in. Neither, during the long silence after leaving Hobart on 1 December of the previous year, had Douglas been able to hear any news of Paquita or his ailing parents.

On Sunday 6 October the men at the Winter Quarters ate a 'special plum pudding made by Paquita . . . It was voted the best of all we have had and I certainly thought so too', wrote Douglas in his diary.

From mid-October, when the first penguin waddled ashore in Commonwealth Bay, there were intermittent signs of spring—such as spring is in those latitudes. Just as intermittently, the relentless blizzards would return, thwarting once again the plans to leave the Hut.

But Douglas's spirit was picking up, watching, as Lennard Bickel so eloquently describes, 'the chill world come to life again':

> Thousands of penguins struggled out of the heaving bay to find their old haunts . . . the seals lunged and lolloped onto the rocks . . . the surge of returning life lifted his heart and gave a glow to the last weeks of waiting for the first marches of exploration. He found in himself a deep affinity and a lasting affection for all forms of life that struggled hardily against a savage environment . . .[2]

Almost a year after he'd said his farewells to Paquita, Douglas wrote again to his love.

[1] See D. Mawson, *The Home of the Blizzard*; Jacka & Jacka (eds), *Mawson's Antarctic Diaries*; P. Mawson, *Mawson of the Antarctic*.
[2] Bickel, *This Accursed Land*, p. 63.

Winter Quarters
Commonwealth Bay
Adelie Land
9 November 1912

My very dear Paquita

This is the first occasion since landing in Antarctica that I have addressed myself to you in writing, though daily a warm glow of life feels to have crept in to me coming from the far distant civilised world, and of course it can be from none but *you*.

I have concluded, once again, that it is nice to be in love, even here in Antarctica with the focus of the heart strings far far away.

Here in a primitive world, in its most rigid aspect with an expanse of tempest tossed ocean between come warm messages straight from your heart, born in earlier days when you and I were together.

Although not writing, daily I feel to hold communion with you in dreaming reverie of all our former happiness. You see I have been reaping comfort for having spoken to you on that quiet dark evening at El Rincon, now nearly 2 years ago. (How the fime flies!) How I sometimes think that you are paying compound interest on my life loaned to the 'Wild'.

The non-fulfilment of the expected wireless messages will have been attended with anxiety and then the *Aurora* might have been caught in the pack or sunk after leaving us. All is so doubtful that I know my true love must have been harrassed by a multitude of doubts and I hope never to incurr such in future. Indeed, I even look forward to making up for this offence.

Down here, things are different: I feel that you are in the best keeping and I only hope that you are passing the time as pleasantly as may be.

Dear Paquita I am writing this note in case anything may happen which will prevent me reaching you as soon as the mail from here, which is expected to be picked up next January. So many things may intervene for truly one lives but from day to day here and then our sledging journey is about to commence.

How terribly disappointing this land has been. Our only consolation is that we feel that everything has been done that could be done and that on account of the rigour of the climate the information that we have obtained will be of special value.

Since the ship left in Jan. last, we have had but a few days of calm weather and the wind has blown with such terrific force as to completely eclipse anything previously known elsewhere in the world. Some of the men have done such remarkably good work in the hurricane wind as to call for admiration from anybody. I trust & hope that better conditions will be given us during the coming weeks.

10 November 1912

The weather is fine this morning though the wind still blows—we shall get away in an hours time. I have two good companions Dr Mertz and Lieut Ninnis. It is unlikely that any harm will happen to us but should I not return to you in Australia please know that I truly loved you from an admiration of *your spirit*. And should we meet afterwards under other circumstances please know and love me as a brother.

In case of my non-return my total assets come to somewhere about £2000 including of course salary at the rate of £400 per annum from the expedition which is paid in lieu of my University salary. Accounts in the Bank of Australasia. I have told you of other things including all my photos and private belongings at the University. Take what you want of all and if any remains you can give to my Mother if alive in lieu to my brother.

I must be closing now as the others are waiting—give my admiration and love to all the Delprats, each one separately—I am writing only to your Mother and Brother.

Good Bye my Darling may God keep and Bless and Protect you.

    Your Douglas

By a later oversight on the part of Captain Davis, Douglas's composite letters of 9 and 10 November were not delivered to Paquita on the *Aurora*'s return to Australia in mid-March 1913. Douglas was not to know of this for another thirteen months, nor how progressively heartbroken Paquita was that Captain Davis had brought nothing for her from her stranded fiancé. In that everlasting silence, the love bonds became stretched to the limit during 1913: a pivotal year, not only for the physical survival of Douglas, but also for the survival of Paquita's support and trust.

November 1912 to November 1913

MUCH HAS BEEN written of the Far Eastern sledging journey on which Mawson, Mertz and Ninnis—with teams of huskies—set out from Commonwealth Bay in November 1912. While I will not elaborate on what has become known as one of the most extraordinary epics of polar survival, a brief account may enhance the reader's understanding of the letters which follow. The experience was to affect the remainder of Douglas's life and to cause Paquita an additional year's wait for her man to return from Antarctica.

Five weeks and more than five hundred hard-won miles out from the Hut, Lieutenant Belgrave Ninnis, together with the best dogs and a sledge packed with all the dog food and most of what Douglas described as 'man food', disappeared down a deep crevasse. 'May God help us', Douglas wrote in his diary.

With their resources thus shattered, Douglas and his remaining companion turned back, surviving on drastically reduced rations augmented by the unpalatable meat from their starving dogs. The strength of the two men quickly waned. they were besought by crippling ailments and injuries. Their escalating privations included the loss of 'all skin of legs and private parts',[1] the suspected cause of which was not to be recognised for another sixty years—they were being poisoned by eating the huskies' livers.[2]

Despite debilitaing injuries and ailments making their progresss slow and painful, they forced themselves on for another three weeks until Xavier Mertz's condition was so low that he could no longer even rise from his sleeping bag. In the early hours of 8 January 1913, he died.

With the deaths of both his companions, Douglas faced the ever-worsening situation in conditions barely conceivable even to him, with his experience of Antarctica's rigours. Alone, and in extremely low physical condition, he still faced nearly a hundred miles of icy plateau and jagged glacier. Thoughts of Paquita, his supporters and the other expeditioners helped him to rally. In his diary of 8 January he pledged to continue to do his utmost for their sake.

In an emaciated state, with the soles of his feet detached, his spleen and liver swollen, his balance disrupted, his food supplies and equipment profoundly inadequate, and in extreme weather conditions, Douglas staggered on.

Three weeks later, on 29 January, he found a cairn, a note, and a bag of food left for him by a search party which had come out from Commonwealth Bay. Inside the waterproof bag was a red Paquita bag, containing the cache of food. This welcoming symbol of better days, a glimmer of hope for warmth, restored health and companionship, greatly touched Douglas. The bag, stitched by Paquita in Adelaide during 1911, seemed a shining emissary of her love for him. He hugged the red Paquita bag and whispered her name, remembering her patience and her support of the expedition.

But it was another ten days before he reached the Hut at Main Base.

Captain Davis had meanwhile returned in the *Aurora*—as arranged—in mid-January. His journals reflect the loyal seafarer's deep concern during his three weeks of standing off Winter Quarters in heavy weather, waiting for all the expeditioners to return from their summer sledging journeys. The non-appearance of Douglas and his companions was particularly worrying to Davis, but he was assured by the other men that Mawson's party had 'plenty of grub, that is a great comfort'. Little did they dream that the food lay hundreds of feet down a crevasse, along with young Ninnis.

Douglas had left a note in the Hut delegating command of the Expedition to Captain Davis 'in the event of the failure of my own party to return' by 15 January. As that month drew to a close, extra coal and provisions were unloaded from *Aurora*; additional penguin and seal meat was stored at the Hut; and, under the supervision of the ship's Mate, improvements were made to the wireless equipment and mast. All these preparations proceeded on the growing assumption that it may become necessary for a relief party to be left behind—yet with all the men still hoping that the three missing members of the AAE would arrive in time to sail off with them. In his journal on 31 January Davis wrote, 'What can we do to help him? . . . God help a man adrift in this cursed country'. Three days later Davis was 'absolutely worn out [with] the worry and anxiety'.

Davis was concerned that the season was rapidly closing in and that before he risked the *Aurora* becoming trapped in pack-ice, he needed to collect Wild and his seven companions from the Shackleton Ice Shelf. On the one hand he knew that the provisions of Wild and his companions would be running low; on the other, that he could leave a relief party at Commonwealth Bay secure for another year. Davis therefore reluctantly left the vicinity of Winter Quarters on 8 February 1913.

Later on the very same day, Douglas staggered back to the Hut, to be met by the small party left behind by Davis. This included the Expedition's young medico, Archie McLean, and the party's leader, chosen by Davis, Cecil Madigan. Madigan had deferred a Rhodes Scholarship to Oxford University to take part in the AAE, and—like Mawson—had left a fiancée in Australia. Now he was committed for a further period of separation from both his fiancée and Oxford, but graciously wrote a memo to Captain Davis thanking him 'for having done all in your power for our comfort for another year in the Antarctic'.

While Douglas was making his first physical and emotional adjustments to rest, warmth, nourishment and human fellowship, the world was learning of the Antarctic deaths of Captain Robert Scott and his four companions. Simultaneously, Sidney Jeffryes the 27-year-old relief wireless operator left by Davis at Commonwealth Bay, was trying to contact the outside world to tell of Dr Mawson's amazing survival and of the deaths of his companions. A particularly active aurora australis produced such an electric disturbance at that time, however, that any satisfactory reception of the many Morse code signals being sent from Adelie Land was rendered impossible. It was 24 February before Macquarie Island had the full story to relay to the world.

During the next fortnight, Douglas used the available 'air time' to send and receive many official messages. He longed to make contact with Paquita, but Paquita was still at sea on the voyage home to Australia. On 7 March, Douglas sent a wireless message to her father in Adelaide, who replied:

HEARTFELT SYMPATHY WITH YOUR TERRIBLE EXPERIENCE. TAKE GOOD CARE OF YOURSELF. ALL MY FAMILY WELL. ARRIVING ADELAIDE END MONTH.

On Paquita's arrival at Tynte Street on 1 April, there was an urgent telegram waiting for her, extracted by Eitel from a longer (business) wireless message from Douglas:

DEEPLY REGRET DELAY ONLY JUST MANAGED REACH HUT EFFECTS NOW GONE BUT LOST MY HAIR YOU ARE FREE TO CONSIDER YOUR CONTRACT BUT TRUST YOU WILL NOT ABANDON YOUR SECOND HAND DOUGLAS

Without hesitation, Paquita telegraphed back to Eitel to send to Douglas by wireless:

DEEPLY THANKFUL YOU ARE SAFE WARMEST WELCOME AWAITING YOUR HUNTERS RETURN REGARDING CONTRACT SAME AS EVER ONLY MORE SO THOUGHTS ALWAYS WITH YOU ALL WELL HERE

MONTHS SOON PASS TAKE THINGS EASIER THIS WINTER SPEAK AS OFTEN AS POSSIBLE

To which Douglas replied

GREATLY ENJOYED YOUR MESSAGE GRATIFIED YOU WELL GLAD LETTERS PARCEL LOVE TO SELF AND FAMILY MCLEAN PRODUCING QUITE GOOD RESULTS NEW HAIR RESTORER AM VERY FIT NOW

Of course, Douglas was being stoic, optimistic and protective of Paquita when he proclaimed himself very fit. It took many more months of Archie McLean's nursing and Douglas's sometimes over-stretched patience for him to regain anything like his former health.

On 3 April, Edward Stirling, Professor of Physiology at the University of Adelaide and Director of the South Australian Museum, included the words 'PAQUITA LOOKING WELL BUSY UNPACKING' in a wireless message to Commonwealth Bay. Twelve days later, Paquita telegraphed Eitel to include in one of his messages:

NEXT MARCH VERY FAR IMPOSSIBLE DO WITHOUT YOU MUCH LONGER FAMILY SENDS LOVE TAKE CARE SELF GREETINGS TO COMRADES

When the weather conditions allowed wireless transmission from Adelie Land, an occasional few words for Paquita attached to the end of official messages was some solace, but real communication between the lovers had to continue by undeliverable letters. The infrequent and irregular Morse code messages passing through many hands—not to mention the world's press—allowed for little real expression of their feelings.

[1] Jacka and Jacka (eds), *Mawson's Antarctic Diaries*, p. 158.
[2] Cleland and Southcott, 'Hypervitaminosis A'; Bickel, *This Accursed Land*, pp. 109–12. Some polar authorities are now questioning the Vitamin A theory; See Ayres, *Mawson*, pp. 80, 81.

Winter Quarters
15 April 1913

My more than Dutiful,

There are many reasons why you, dearest, occupy my mind more this year than last. Don't imagine that, even then, you were not a daily visitor pushing your way in amongst the intricacies of my more immediate thoughts.

Well, to begin with, there is the consideration that last year everything was prearranged and one had resigned ones self to the inevitable, even ere final caresses sealed the interval to emotional oblivion (if indeed such a thing can be recognised).

How different this year—this unexpected happening having put a deeply grudged interval upon our happiness.

But little of scientific work remaining to be done in this locality, time hangs heavily. Imaginative fancyings carry me to a little snuggery presided over by a gentle fairy—where all is peace and joy and quiet—an antidote to the one-half life and a balm to a hurt mind.

There in the radiant glow of a home fireside; spanning the span of life, tended by and tending the most exquisite of companions.

Then there is the receipt of your delightful letters[1]—on the other hand, a knowledge that I have introduced into your life of baby-bliss, anxieties and uncertainties which must haunt each day with fretful forebodings.

How sweet your Mother to lull your anxious moments with suggestion of telepathic premonition. Indeed this subject has given me food for thought of late. My good Father, who by the grace of God is now partaking of a kindlier sphere,[2] came to me in a vivid dream whilst sledging. May it not have been in connection with his demise which happened about that time? Whilst on this subject I am sad to relate that news from Campbelltown reports Mother's condition as continually worse and worse.

Coming back to the old subject—*You*. I have stated some of the reasons why—but there are others. Your Photo has been housed in a frame which together with your mascot resides above my pillow. In some of these amateur photo scraps you showed a keen appreciation of my love.

Now this leads up to the wireless which I sent you. It seemed to me that, after all, I was of most ordinary importance and with the delay and indeed uncertainty of return was selfishly occupying your love against a more certain and less anxious future. Added to this, a fundamental prompter, was the fact that I had been very roughly used on the treck towards the hut, as the gateway to Earthly futurity. Though this demonstrated my physical ability at the time it was a great shock to my entire system and the effects of it left me more than ever convinced that I did not merit your appropriation.

Time has largely healed my scars and now perhaps I would be more selfish in consideration of yourself. It may have appeared to you in your love, a very sordid matter—the theme of my bald statement—nevertheless I would not have you marry me for consideration of aught but true love and not then if gnarled by any seeming imperfection.

Aye—a dreamer am I in this desolation lulled by the brooding solitudes of the Polar lands to roaming in the empyrean of mind.

It is now midnight—and where are you?—and where am I? and where might we not have been?

Well my love—Goodnight! and here is a kiss to redden your more than ruby lips.

Douglas

[1] Although, on his return to Australia in March 1913, Captain Davis failed to deliver Douglas's letters of 9 and 10 November 1912 to Paquita, he did leave in the Hut at Commonwealth Bay Paquita's four letters to Douglas of February to October 1912.

[2] Robert Mawson died at Campbelltown (NSW) on 10 November 1912.

Tynte Street
[North Adelaide]
21 April [1913]

My Douglas, Oh my dear, dear Man.

If only I could come with this letter. Oh that terrible journey all by yourself if only I could have had it instead of you—did my love help you then? It is worth less than nothing if it didn't. Dear you *have* had a hard struggle down there—thank heaven I have you still. Douglas I can't find words for it all—my heart is full of thankfulness. Miraculously spared to do great things—a life is not spared like that for nothing & I shall help you do them. Poor poor Ninnis and Mertz. Correll tells me they were so much liked. Mertz was not so wiry as you. They both knew the risks they took. Thank heaven, their deaths were not due to any thing that could have been prevented. And you are safe and resting now my love.

Oh Douglas how I long for you! All these eighteen months have you been constantly in my thoughts as you are for ever in my heart. And now more than ever. I have heard all I could about the hut & your life there & picture myself with you helping—loving you. Oh how I hope my love helped you on that journey that it upheld & supported you as I should have had I been there. The disappointment of your non-return counts as nothing against the gratefulness that you are safe & well. And had you not had wireless we should not have known for a year whether you were safe—oh it could have been *ever* so much worse.

Douglas I have grown so much older in these months & I came back from Europe with my head full of things to discuss with you. And now everything has to wait. I feel the need of you now—am at a sort of turning point & feel it myself. I *do* want you so badly. When I look back at myself [at] the time you left it seems a different person— so young & silly. How could you love me? I must have hurt you, my love, often but believe me & I know you do it was only in ignorance. I loved you then as a girl who knows *nothing at all* of life & now—as a woman. Oh we shall be so happy. I am not afraid of anything now (did I hear you say cows?) You are coming back to a very much warmer person than you left. And oh there is so much to say about these months of separation. Every night I hold imaginary conversations with you. I never understood when you said that at 25 one loved more than at 20. I thought it an impossibility to love more than I did & remember telling you so! I shall make you as happy as ever I can when you return to me, Darling *darling*.

Everyone is very kind to me. We heard the news at Aden—you can imagine how I felt how we all felt. But I'm missing you now more than on the boat with so many distractions. Prof. Henderson[1] is most kind and has taken us to the theatre & to see Scotts pictures & brings everyone here connected with the expedition. Correll has been twice & Capt. Davis. I like him. Oh for next March. Capt says you might come into Adelaide.

We are moving our domicile to Melbourne in a few months but Mother & I will be waiting for you wherever you land (except Fremantle, I can get no one to come through the Bight!).[2] So come to Adelaide & happiness in March. I may be able to last till then. I sincerely hope so but not a moment longer can I wait. I want you I *want* you & hope the longing is mutual. Just think if you hadn't told me before you left!

We have not yet got into the usual run of things & so I feel the emptiness of Adelaide without you. We all feel strange here after Europe & Melbourne will be much better

This Everlasting Silence

for Liesbeth and the doctors [Lica and Mary] when they come. Adelaide seems a small village[3] & the working class is worse than ever. Its quite different when one has a house & so but for the girls it is very slow here & they feel the lack of music & movement. We all do.

I hope you'll think my voice improved. I sing a lot now & love it more than ever. We want Lica & Mary to come out at the end of the year but I'll tell you later of our movements. Theo & Rene might be here for Xmas. I'll see the *Aurora* in Melbourne.

Jessie says I may come & stay there this year some time.[4] I wish it had been possible to do so before we left Australia. Dearest you know how I feel for you that your father will not see your return. I do hope your mother will be better when you come. She has not yet her memory back.

Dearie I was glad to get the wirelesses & look out for more every day. But they are rather unsatisfactory. I want a letter telling me how you felt all those three weeks & if you are really alright now & how the year has been. Correll tells me the harmony was good amongst the men. Did they worry you or treat you well? They all have a jolly good opinion of you & I think you'll be accorded a very warm welcome. You cook the best they say—a good lookout for me in this domestic-less country.

I won't write more now, though if you feel as I do you won't ever have enough!

Mother is more of an angel than ever. She is very tired just now. Lica has been a little ill but is better now. We are all well now. What happy times we shall have when

November 1912 to November 1913

the family circle is complete & the best & dearest—for that you are dearer than anyone—is back also. One loves one's parents and family in quite a different way. There is no one who could love you more (& this time it is true) than *your own* longing & loving mate. I want just to feel your arms—your lips. Oh dear! I do want you my lover my dearest and best of men.

    Your very own
    Paquita

[1] Professor Henderson had written a warm and kindly letter to Paquita on 17 March 1913, possibly addressing it to the ship's agent in Perth:

> Dear Miss Delprat, Mawson is safe and well. Let that sink into your mind before I tell you anything else. . . . You are engaged to a man who has proved himself a hero of the toughest fibre . . . his return journey as fine a record of endurance as ever was put up in Arctic or Antarctic travels . . . You will be disappointed—bitterly so, but I have no hesitation in telling you that you have no cause now for alarm and very good cause for deep and lasting gratitude.

[2] Before the first transcontinental trains operated in 1917, the only means of transport to Perth was by sea. The seasickness suffered by most passengers through the Great Australian Bight was a discouraging and dreaded factor.

[3] The population of Adelaide in 1913 was 201 000; Melbourne's was 651 000.

[4] Dr Will Mawson's wife, also known as Dally.

That letter of Paquita's contains the first hint of a hidden theme which now runs through her letters to Douglas, namely her sexual innocence. Much later in life, Paquita confided to a friend that, as her father had forbidden his family to acquaint her with the facts-of-life, she had been very afraid that some passionate kisses from Douglas on the eve of his departure had rendered her pregnant. When her medico sisters had learnt of her distress, probably during Paquita's stay in Europe during 1912, they had reassured their young sister, gently educating her in the true facts. Despite a burgeoning of female awareness by this time (connected to the movement for birth control) it was not uncommon for 'protected' girls such as Paquita to be innocent of the biological requisites of reproduction. Even a private school as progressive as Tormore House would not have taught anything about the subject at that time—in fact, such enlightenment was still half a century away.

Winter Quarters
5 May 1913[1]

This is May 5th—It is evening—I am on night watch—all have retired and I am alone. But no! I am never alone now for if the immediate present does not occupy my mind you are there having little talks with me.

A confusion of local events—a hazy uncertainty—then dream-like appears before me the same haughty youthfulness, the same loving self, the tender lips, and more—and lo, it is you. Many little incidences we have had lately, you and I in dreamland. Have you heard them?

I am now running into my 32nd year, 31 is left behind. What has fortune in store for this 32nd year? Surely I can predict a great event never to be forgotten, as the saying is—our meeting in $31\frac{3}{4}$ Anno Douglii. After that I hope we shan't ever stray very far apart, for life is not so long that we shall ever tire of each other. That is if you do not expect too much from me. I know I have all the faults of everybody else and a few original ones besides—so do not think you are about to marry a saint. Only a very ordinary person who is just as likely to put his foot in it as not.

Did you read *Patchwork Papers*? I liked it, that sort of imaginative literature appeals to me. Do you know that had I not lived in the 20th century I might have been something very different, a crusader or a Buccaneer. I am glad that my lot has been cast in University work, with expeditions to let off superfluous steam. Don't have any apprehensions that such will happen again, for you will be somewhat in the position of the engineer, and a *good* engineer can always regulate steam pressure.

Coming back to very *earthly* things, I have to wash some clothes tonight, so I must be off. But not to hasty— I will close my eyes while you impress me with one more kiss, Darling, like you use to—long and sweet————
 Now Good Night
 Douglas

[1] The somewhat subdued tone of this letter, with no opening salutation, could reflect the fact that it was Douglas's birthday. Although his diary for 5 May records a celebratory dinner, an aurora had completely eclipsed any wireless reception. A combined message of birthday greetings, despatched by Eitel on behalf of Paquita, Professor David and himself, was therefore not received until a week later. This second birthday away from Paquita may have caused Douglas to question his dedication to science at the expense of personal fulfilment.

The Patchwork Papers, mentioned in this letter, had been published in London in 1910. Compiled by E. Temple Thurston, the book was a miscellany of stories and essays of interest to both men and women. Some had an early feminist flavour.

Douglas had bought the book in Hobart and sent it back to Paquita from Macquarie Island. Uncharacteristically, she neither thanks Douglas for the book nor comments on its contents. Perhaps, between her departure for Europe early in 1912 and the family's move to Melbourne in mid-1913, *The Patchwork Papers* was either lost or overlooked. Or was it borrowed by another member of the family, who then forgot to return it to Paquita?

The book obviously made a considerable impression upon Douglas. He first mentioned it in his letter of 10 December 1911, then again on 5 May 1913, and also on 1 July 1913—a year and a half after despatching it to Paquita.

During his second winter at Commonwealth Bay, Douglas wrote more frequently to Paquita, despite the full realisation that his letters could not be received by her until the *Aurora* once again returned to Australia in 1914. On re-reading his letters nearly fifty years later, Paquita wrote: 'Douglas . . . had wanted to give me a record of all he was doing and thinking, as a kind of solace if anything happened to him'.

He was all too conscious that, had the Far Eastern sledging journey been fatal for him—as it so nearly was— there would have been so little for Paquita that he had written for or to her after 19 January 1912. If he had a sense of guilt about his omissions of 1912, he certainly made amends during the latter part of 1913 even though Paquita, thousands of miles to the north, was unable to receive his output.

The Hut
1 June 1913

Dearie,

I am just going to give myself more joy, wandering away with you for another short hour.

It is many days since last writing; though, and you may have felt it too, our verbal chats have been often enough. Well there has been much writing on topics for the *Blizzard*[1] and other things, besides there have been some reminders that I am not altogether restored to my old-time health. All is well now and I trust everything will be so for the future.

What shall we talk about this evening? It is so unfair my choosing the subject—and after all you may not be interested.

I am tempted to make you the subject but, on reflection, can see that that would be least of all interesting to you. While to me—well you know. Perhaps I could tell you some points you little think I appreciate— on the other hand you have certainly a magnificent opinion of me—or is it that innate quality, lovable above all others, of full womanhood which drives through the aureole of surrounding circumstances and leaves the woman standing on a pinnacle of isolation, pinned to the body and soul of the fortunate man? Perhaps it is the fullest spelling of the word love.

This Everlasting Silence

This is a bigger subject than most ever imagine. In its complexities, its many avenues each leading out into a jungle of subconscious emotions, it will offer an unfailing subject for analysis and enquiry for ages to come.

It is quite on the cards that as evolution proceeds, the human will develop this now, for the most part, latent faculty to an extraordinary degree—the merest shelter of a savage today into a splendid temple in the future.

Love is unknown in the lower animals, hardly perceivable even in mammals except in human beings. Even of us it is doubtful if true love in the ethical sense, ever enters to any appreciable degree one half of the civilised peoples and much less must it be the case in more savage races. Love in them is merely passion and instinct.

What more true picture can we have of God, the ultimate issue of all things, than the personification of love—Is it that we have in us just so much of God as we have acquired of love?

If all the professing Christians of today would only recognise this, how different would their actions be towards one another. How much more tolerant and forgiving would we all be.

This is Sunday evening but this is not a sermon, dear. It is an issue that comes to all thinking people now a day, when such revolutions are going on especially in the status of woman. What will be the outcome of it all? Will the altered relations of men and women affect love between the sexes? Will it heighten or diminish the probability of love?

It certainly will not eliminate passion, though it is likely to be a cooling influence. It will certainly diminish the number of marriages probably acting by diminution of the percentage of passion marriages and no doubt to some extent diminishing also those where love is the more fundamental bond.

Human beings like other animals can slowly evolve into entirely unlike forms with diverse aims in existence—this possibility and natural process is shown by the variety in genera throughout the whole scale of animate life. But the process is slow, not the evolution of a day nor yet of centuries.

It so happens that man is a combination of certain qualities woman of others. There is a fundamental dissimilarity just as surely as there is similarity. In the biologists classification the female represents the passive vegetative state the male is the active animal state. This generalisation is true of woman and man no matter how much the *new woman* may think to the contrary.

The intellect of woman in no wise negates the more fundamental atribute of womanhood.

Woman has her sphere—man has his—between them is the whole future of the race. How far the one overlaps the other is what should be occupying the minds of the directors of the legislature of to-day.

Christianity more than any other factor has altered the status of woman—and rightly so. It has suppressed the strong and exulted the weak. But by a prolonged streaming in that direction, will it be the power to overthrow the established order of things? The great cataclysm which the future has surely in store, reading by the light of the dim past. What was it that replaced the

This Everlasting Silence

fish of the Palaeozoic Period by Reptiles in the Mesozoic—and in their turn by Mammals in the Cainozoic? What is it that will overthrow the dominion of the genus homo? Is it woman with whom in such a degree lies the fate of the man of tomorrow.

In the meantime will we pass into a condition of matriarchy? Is it not likely that the hordes of Asia, unaffected by these changes may override roughshod the dwindling population of future Europe—and our proud Empires of today be a thing forgotten as the regencies of Egypt and Babylon which also did fall to the insidious germ of social change.

Now I am getting side-tracked—so much depends upon love that it is easy to wander into one of the issues, leaving the fountain head unexplored.

Don't mind me rambling on, it is so easy to write of love when you are so far away and your love so near.

What is it that is bruited about amongst society as love? A simple enough word, but is the colloquial usage really so simple? Are not several distinct attributes commonly paraded before the world as love? The three most common are perhaps (a) Instinct, say between mother and child (b) Passion (these are both really subdivision of the one). (c) Then there is love—a state of bond beween the ego of one person and that of another. This is of course true love.

Surely it is the quintessence of the latter that is between us two—though I do not deny my full share of passion when you are about, any less than I breathe and eat.

How more than human will it be to breathe and eat with you!

November 1912 to November 1913

What has made me run on like this? I have entered the subject love with you as presiding deity and raced off goodness knows where—please forgive. When we are together there won't be any side attractions—love will mean nothing but ever so near your heart, as near as you will allow.

I have been writing near the stove and my feet are beautifully warm—so will now go to bed to think of you, perchance to dream.

Don't come to me as ivory—as you sometimes do—I want you warm and—well *you*.

I have kissed this X spot, dreaming that it is near your heart.

> Good night my well beloved———
> Stay! One more X———
> really I must go now
> Douglas

[1] *The Adelie Blizzard*, a magazine produced at the Hut on the Expedition's manual typewriter between April and October 1913. In the tradition of polar exploration, it aimed to relieve winter monotony and depression. *The Adelie Blizzard* was edited by Archie McLean, whose medical care had been so vital, and who in 1914 was to help Douglas Mawson write his book *The Home of the Blizzard*. Issues of the magazine are preserved in the Mawson Collection, Waite Campus of the University of Adelaide.

Winter Quarters
Antarctica
10 June 1913

Night watch again Dearie and I am so wanting you. I wonder what we would talk about were you here in the person. Of course allowing for a preliminary interval in which we did not talk at all—only felt. And how much deeper does one not feel than words give utterance to—with your head resting on my shoulder, or my head on your breast, have it as you will, such floods would pass between us—words are not in it. Good things to hide one's real feelings under when used with a nice skill, but never to express the heart when dumb touch is a competitor.

Ah well, we *can* talk but alas touch must be reserved, let us agree, at compound interest.

Now it has just occurred to me that there is one thing that I wish particularly to speak about—that is the wireless communication or more correctly the reverse.

The top section of the mast was shattered 3 days ago by heavy whirlies. It had held up so bravely that we began to loose anxiety about it—and then to suddenly carry away like that. However we shall see what can be done to get in touch again.

Of the bad fruits of this mishap the thing I feel most badly about relates to you. In short I should have sent you more messages. Let me tell you the reasons of the paucity.

(1) The auroras, which became more frequent as the winter advanced, interfered daily to a greater extent with the sending. Days passed without being able to get a word through.

(2) There were a great many business messages I was anxious to get through and these became more and more congested with the increased difficulties in sending.

(3) Until the business messages were got off, it was diplomatic of me to send few private messages for the

others here were clamouring to send some to their loved ones also.

Now that the chance has gone, I would throw everything to the winds to get a word in to prevent you worrying.

There were, I think, 4 communications in all to you—one of my escape—one in gratitude of your love—one sending *much* love—and another thanks for birthday greetings.

A miserable paucity to keep my large hearted love free from anxiety for a year.

I hope and hope that you are not worrying. Have often wondered whether Captain Davis posted several letters I had written and locked up in my box—one of them to you. Others to Will, Mother, Father and Professor David. I hear that, not expecting my return, he had opened the box and searched it looking for orders before leaving here.

I remember just before sledging, writing those short notes and never thinking that they would be posted unless I was not returning. Never mind—I don't suppose they contain anything I should be sorry for. Had I known what lay ahead how I should have devoted myself to writing to you last year.

Now Dearie, I trust you are just your old self tonight, and no worry—soft slumbers, sweet dreams, the warm heart I can be sure of.

What would I not give now to slip in to you—asleep—and steal the honey from your lips, sweeter than all else—and then perchance dame nature, doing her work well, would have you dream that I was well and happy to the seventh heaven of delight.

Ah! Dear love those lips are precious, be sure I do not value them too lightly—precious, why so precious?—why? That is our search.

It is nearly 4 o'clock and I must not keep you awake any longer. Before I go let me tell you a parable.

Once upon a time there was a beautiful princess and a very ordinary man. The princess lived in a far off country in a big town, and the poor man was very lonely in a desert place but he used to get his happiness by thinking of the princess.

And always when he thought of her a feeling came to him just as if he would like to burst all the way down past his heart—and to open out and envelop the princess—to sort of run together all into one—learned people would say to mingle his entity with hers.

Surely this is love.
> Kisses and Good Night
> Douglas

Winter Quarters
17 June [1913]

Paquita

The gramaphone has just been running for an hour: you have no idea how beautiful it is down here, in the absence of anything better. How it makes me long to be back to a wider range of music—how I want to hear you sing again. If I could steal on you, unbeknown. 'The Blue Alsacian Mountains' is on now—it makes me think of you. We are really very well off for records and they do, indeed, help to fill in a gap. I wonder how you are feeling in my absence—If you only knew how much I love you, surely it would help—and then it won't be long now.

> In the year that's come and gone, dear, we wore a tether
> All of gracious words and thoughts, binding two
>   together.
> In the year that's coming on with its wealth of roses
> We shall weave it stronger yet, ere the cycle closes.

There are so many things we should be talking over, considering that we are very nearly on the eve of that auspicious occasion when we shall jointly 'light our lamp and wait life's mysterious tomorrow'. What an important event. And the intervening days are slipping by, unfortunately not at their accustomed speed, but as ever 'time and the hour runs through the roughest day' and our day *will* come.

I am certain to be very busy on arrival in Australia for, whilst the iron is hot I must push for funds to pay off all salaries, the money for which now appears to be very short owing to this unexpected extension of the expedition. It is my present intention, however, to take say 2 weeks in retreat, so to speak, in Adelaide before lecturing—this will give the opportunity for us to talk over everything and also I can go over the accounts and plan the finale—a programme which will occupy about one year, but really need be no bar to our union, as I hope it may not.

Let me diverge to instruct you in all that is to be done on my return—curtly summed up it is the clearing off of all debt. The way that this will be done can be stated under the following broad heads: (a) Lecturing in Australia & N. Zealand. (b) Publication of a book in England, also magazine articles etc. (c) Lecturing in England, also possibly a few on the Continent and in America. (d) If the debt is too great to be cleared off in this way, an appeal to the Government for the same, this if necessary will of course be made immediately on return.

I say 'if necessary' because I am really entirely ignorant of the state of the funds, it not being a wise thing to get details by telegraph, and Eitel not leaving a statement when the *Aurora* left as they had entirely given me up.[1] The money in hand should have at least been enough without taking account of the added programme this year—which will mean £5000 more. From hints received, however, I believe they spent more than they should have last year and that the debt will be greater. They should have made a strong appeal to the governments *now* whilst we are down here and got every penny necessary, which I believe I would have no difficulty in doing had I been there—however. I have discovered once again that an expedition lies in one man only and I can't expect to hear much to my advantage on return.

It looks as if all the money forthcoming from the book, lectures etc will have to go to pay off debts. I had always anticipated this money would be available to bring out the scientific results and part to my private account. No other expedition leader has thought of devoting any of the returns, after paying expenses, to aught else than private account. It was a big concession that I had intended seeing the publication of the scientific results well footed. You note that apart from my position as leader, I was the originator, promoter and took a great financial risk to my own private account at the commencement if not at all times. Now I cannot promise you a cent from lectures or book though the gross returns will aggregate from £7,000 to £10,000.

My private banking account should, in March 1914, be quite £2,000. If government money is received to pay off salaries it will be £2500—count on £2000. Beyond that a chance of more, somewhat more certain of results than a ticket in Tattersalls. My Adelaide University salary is now £500 per annum and it can't be many years before it is £800.

Now you will agree that this is enough to marry on, the question is when is it to be.

Now there is no doubt but that I shall be much happier under your domestic eye, so everything else being equal, the sooner the better—Eh?

Now Adelaide University has been very good to me in the matter of leave, so that I am anxious to follow their wishes as much as possible. They will want me to go on with the lecturing in 1914 and it may be well nigh impossible for me to get away to England and lecture in Australia. But do it I must, even though it hurts them (I think I am too valuable for them to abandon me). To get away I will promise to be sure to return and lecture the following year [1915].

Admitting that they give me leave for 1914 then it appears that the following course will be the best.

Arrive Australia 1st March—in retreat until 15th March—sail for N. Zealand & lecture & attack the Govt. for money—arrive Sydney Apl 7th & lecture Australia until end May—Leave for England having already got manuscript for the book about ready—Busy with seeing the book through the press until end of August—Commence lectures in November opening before Geog. Soc.[2]—Give about 30 lects in England by end Dec. Give few to Geog Socs. on the Continent in Jan 1915—Return to Australia, probably by way of America giving a few lectures there. Returning to Australia probably at the end of March—possibly not until the end of May. The American lecturing might be done in October & September 1914.

Now where do *we* come in—It seems to me that we should not miss the opportunity for a honeymoon. Perhaps that would be an additional motive for asking for leave from the University. I *would* like to see a little of the world *with* you. With regard to the expense, even if things go badly I shall debit my own expenses to the expedition account, also the £250 necessary to provide a substitute at the University. Should the expedition funds go well, I can debit our joint expenses up against the returns from the lectures etc.

Should the added expenses of having my wife with me have to be footed privately it would amount to £400. In my absence I shall get only £250 from my University lectureship, so that by the end of the year we should have drawn on the principle to the extent of £150. Perhaps you will agree with me that it is worth it. Very likely we could

November 1912 to November 1913

effect a saving by buying some of the embellishments of our bower in England.

To get married and furnish a house nicely but not sumptuously will cost about £650 (including marriage expenses). N'est ce pas? With the £150 mentioned above the expenditure from principle by March 1915 including furnishing of house would be £800. Leaving only £1200 in the bank and a sure £500 per annum subject to rises very soon.

I anticipate it will cost £400 per annum to live in the above furnished house (all expenses). So we should be £100 per annum on the safe side and say $3\frac{1}{2}$% interest on £1,200.

By the way, my present life policy is only £500 with bonuses payable on death or at 50 years (I think it is) of age. I took out another one when we got engaged but as they heard I was going to Antarctica they made me take it back.

Recapitulating Our position say in March 1915, having married at the end of May or early June 1914 and toured as stated and furnished to the extent named and contracting annual debts to the extent of £400 per annum would be—

(1) House furnished

(2) Income from University lectureship £100 over expenditure (and good prospect of speedy advance)

(3) In the bank £1,200 drawing interest at $3\frac{1}{2}$%

(4) A life policy equivalent to about £650 on death, increasing slightly every year.

What a long rigmarole—you will say, but I hope you are interested.[3] Perhaps you will not agree with the figures.

This Everlasting Silence

But I really think the expenditure will be of the order named, only time will decide.

It will be love in a cottage, but Oh such a nice one with you presiding—Indeed a home will be very nice—I *have* been a wanderer and home *will* be novel.

In the last 13 years, on return, I will have spent quite 4½ years out of a proper house—of this 9½ months sleeping in tents on snow and 14 months camped in tents in the New Hebrides and elsewhere. For the rest, what is a boarding house and hotel existence to *home*—with you as the presiding deity there will be 'no place like home'.

This is rather a business letter but I seem to have been in the mood for it and down it goes.

    Good night, my own *Darling*
    Douglas

---

[1] Conrad Eitel, secretary to the AAE, had sailed on *Aurora* as a supernumary when Davis had returned to collect the expeditioners late in 1912.

[2] Royal Geographical Society.

[3] While Douglas may well have under-estimated the costs of setting up a home, he was ahead of his time in giving his fiancée such a comprehensive summary of his finances in an era when most men of his standing expected their wives to know little of financial matters. The irony is that while Paquita would have appreciated the frankness, the intricacies of the fiscal reckonings may have left her head reeling. Many more years were to elapse before she had much appreciation of budgeting! But as this letter —together with all the others from Antarctica—was not received by her until the *Aurora* brought her Douglas in person, it's unlikely that, in the fervour of excitement, Paquita ever paid much attention to the financial matters it outlined.

November 1912 to November 1913

Now that he was slowly regaining his health, and had attended to many outstanding business matters, Douglas set to work answering Paquita's four letters—written between February and October of the previous year. They had been left at the Hut by Captain Davis while Douglas was still battling the rigours, disappointments and tragedies of the Far Eastern sledging journey.

While much of the news on which he comments was therefore out of date, and while Paquita had matured beyond some of the views expressed, they were—at this time—all that Douglas had from her. That he valued the letters is obvious from the diligence of his conventional 'answering' them, item by item. Although he sometimes writes in the present tense when commenting on occurrences or attitudes of a year or so before, he was intent upon creating for Paquita some record of his thoughts, in case something befell him to prevent their reunion in 1914.

It becomes increasingly obvious in these 1913 letters of Douglas to Paquita that he was trying to put the nightmare of those three months of the Far Eastern sledging journey behind him. Certainly he didn't 'unburden' himself to her, much as we suspect she would have felt for him had he done so.

The weekly additions to this next, long letter reflect the fact that as Douglas's health slowly returned, he took his turn at being nightwatchman, along with the six remaining members of the expedition at Commonwealth Bay. During the previous winter, when there had been seventeen others based at the Hut, Douglas had rostered himself as nighwatchman every eighteenth day.

This Everlasting Silence

Winter Quarters
Antarctica
22 June [1913], 11 pm
(Midwinters Day)

Hooray! Old Sol is now steadily rising towards summer. He started to come back at 8 am, or should have according to calculation. The snow has been falling so dense all day, however, that we have hardly had a glimpse of daylight. By a big dinner helped out with a few bottles of wine, we have suitably punctuated the occasion. This rising of the Sun means a lot to us—It is the visible sign of coming summer and return to our dear ones. *Paquita*, my own Darling, you all in all to me, even so is my ardent longing.

Those four valuable letters from you have been reread on the occasion of each celebration during the past few months. Tonight I have feasted on them again, also your Mother's, Lica's and Mary's, all much appreciated letters.

It has occurred to me that I might answer yours forthwith, taking them in order.

So you liked the voyage: and Colombo! I have a special attraction for Ceylon, or to be more correct, Ceylon has for me—but a few weeks of its tropical steamy heat would suffice.

But the time *will* come when we are so, and I am sure even happier than *your French couple*.

I am glad your acquaintance could meet out consolation for my absence, but I trust it was only partial for I *am sure* your love is deeper than acquaintanceship. But please don't be cruel with him, for he must perforce admire you and then—well don't be cruel, let him down gently.

November 1912 to November 1913

That press notice of a mail in the autumn of 1912 was a hoax like, I dare suppose, many things which have appeared in the papers in the papers in reference to this expedition. Much that is untrue and exaggerated gets into the papers, even regarding expeditions. Editors are not particular, and are often misled.

So you are a 'good sailor'—I am sorry to hear that Mother and Lica were ill, however they are healthy and strong and it won't have hurt them. I don't think that I could ever be sick on those large boats, that is unless complicated with billiousness or something abetting it— though I am sick at times in boats like the *Nimrod* and *Aurora*; but even then, there is so much to do that it takes a back seat.

Dear, regarding my words 'that I would not have time to read etc'—how often have I bitterly rued them— may I never tell so false a lie again. Why tear those letters? If you only knew how valuable is each piece of paper from you.

To bunk now (I will not call it bed, for it smacks of profanation)—more another time.

Tuesday 24 June, 3 am

Night watch and I hie me to your letters—

From your no (2) letter—'Whatever you write I know what you feel'. I am so very glad that you understand so profoundly—that is the root of everything between us. I may be moody today or bad to you unthinkingly tomorrow or in short anything may happen but I cannot see my self ever changing from a warm love, deep rooted, and

This Everlasting Silence

perpetual which will ever thrive in the radiance of your own and, like a flood tide over all the little things which experience teaches us are to be expected.

So you did not go to Spain, but you say 'one plan is fast and sure, we will be in Adelaide before you arrive.' Dearie, why did you not touch wood when writing that.

'It was cold there. I felt as if I was with you'. Let me say in the words of Swift 'You sauce box'. It is likewise so kind of you to remind me that 'you don't' (mind the cold). Now don't be too sure of Lica, you monkey.

A *real Dutch House*—well dearie, if it is as 'comfortable' as *some* Dutch people it will do. Re bed, I would like to be there too, but as am on night watch must perforce sit up until 10 am. What ho! Will I! Won't I! My word wait till you see.

I am really quite sorry you have reverted to Holland again—well go ahead—So you *love* it. Let me tell you I am getting a little envious—sit down Holland! Where do I come in?

So Paquita you have been thinking of me every night —ah, if I *were* only there now. Yes it *is* your warm heart.

Yes, you are right, there is compound interest due, now on two years. Shan't we have a jolly time?

That is certainly a remarkable border 'ducky chickens' —Oh, I say let us have a turkey gosling border for Sunday's tablecloth. I am so pleased you are interested in our furnishing, we *shall* make it the *nicest* home in Adelaide or wherever it happens to be—that is as far as taste goes. Of course in all cases it *will* be the *nicest of all homes*.

November 1912 to November 1913

Ah Dear those lips most rare—they make me feel quite ~~quite~~ besides my self (I only wish it was besides you)

> *Those lips of ruby and Paquiterine,*
> *Far sweeter than the choicest nectarine,*
> *Which, in their sweet convincing diction,*
> *Emphatically assure me love's no fiction.*
> *Would that I could fall upon them now;*
> *For all that's mist, shall double measure yield I vow.*

Bad verse but you can't expect more from your proxy iceberg.

Well now that is the end of No 2 and No 3 must be reserved for another occasion.

1 July 1913, 1 am, night watch

By the way—I surely wrote you from Macquarie Id., and trust you got it—that would be before you left for Europe. —How do you like *Patchwork Papers*—I sent it because I like the feeling it mirrors—perhaps you do not like that class of book—Is it a case of Jack Spratt and his wife? At any rate we will agree to lick the platter clean between us.

So you remembered my birthday—I suppose that 'Old Dutch spoon' was given you to make up for what we are missing in the spooning line. However, I am very pleased that they take notice of my birthday, though soon I shall be getting on to that age when most people like them forgotten.

'I hope you don't mind but I am quite Dutch'—What do you think I take you for. You are not a bit like a Maori, or an Asiatic or anything like that. You may be like a

This Everlasting Silence

Spaniard but I don't know enough of them, I should say your skin would be too fair. An Italian? No. French? No. English? Well just possibly, for you know we are a little bit of everything (the best of all the European nations came over to England, you know: including the Norse, Danes, French hugonots & Dutch dissenters). German? No— Dutch? Well, I suppose you *must be* but what about the dark hair and eyes Eh? Surely not normal in Dutch— however, there must be a large admixture of Mediterranean blood (the origin of all the dark hair in Europe) in the Netherlands, especially in Belgium. As for myself, I am probably a mixture of Celtic, Norse, and Danish with a little of a few others thrown in. Even if I should mind, I suppose I shall have to put up with your being Dutch. Eh? At any rate I will be quite right in calling you 'my old Dutch' though the epithet 'old' will not hold for some time yet.

By the way, I have just been looking things up and find that when Julius Caesar called on your country you were all living in mud huts, in the swamps at the estuary of the Rhine. I wonder what your ancestor of those days of simple fashions was doing at the time? It must have been very distressing for the Dutch women of those days, for they would never have been able to get the mud out of the house, a thing which they are reputed nowadays, never to tire of doing. Those must have been fine, and simple days for housekeepers—not much in the way of clothes to mend—no floors to wash—no fancy cooking—no washing up—for meals just grab a leg of dog or whatever happened to be the 'piece de resistance' and knaw it off hand. I wonder how I would have liked 'my old Dutch' had I met her in those days? At any rate I am quite sure we would have had the nicest mud house in the village.

'My Holland' what about 'My England'

I can just imagine that we are on that little trip together. Won't it be jolly? Very likely I shall have a difficulty to get away from the University for another year on leave. Anticipate it will be all right however. I feel that I should not now throw up that position. First because I am grateful to people who are kind to me: Second there will be much editorial work for 3 years, getting out the expeditions scientific results, which if on top of additional work certain to attend a new office will leave no leisure at all: Thirdly the salary is better than I will get anywhere else in the world for the amount of work that I do: Fourthly, should I elect to remain there, within from 2 to 5 years there should be a professorship of geology at the A.U.[2] and I would get it (I have touched wood). Then if 'we' desire I could probably get a better position elsewhere.

I say! That is a fine idea—a book of ideas—I wonder what it is all about—suppose the ideas don't run on such lines as for instance 'your bringing me breakfast in bed'. I mean of course the other way round, that was a wrong idea and I suppose you have no book for them. Perhaps they are ideas about furniture and things—I trust they are suitable for English people, for I am not very Dutch you know.

So you are collecting house linen—also old brass ('any rags, any bones any bottles today').

So you are glad—So am I, tender one. Always together—Always together. It sounds very nice. I hope

This Everlasting Silence

you won't get tired of me—I am thinking it will be very nice—fancy being together now—me with quite a big set of whiskers—wouldn't it surprise you—in the dark—Eh?

Well I *have* gone and done it—'stopped away another year'. I[t] must have been the 'contract' or something that made me get back thus far because I very nearly did not.

So you are still boosting up the Queen—I should think not—'beautiful'.

So Lica is a duck—I had my suspicions—well I hope that won't disqualify her from visiting our 'dove cot', for a 'duck' in a 'dove cot' would surely be an interesting circumstance.

So you are rubbing up languages—I shall have to raise your salary and make you interpreter.

Re Willie. Every boy is likely to fall in love before he knows the rules of the game—antidote, plenty of outdoor sports and less of women's company.

Leinte is also a Duck—good heavens, whatever has your Mother been doing. I suppose I shall see you all waddling down to meet me at the train.

Yes! You are right. I have been counting myself up on my fingers, and come to the same conclusion each time—I am *quite sure* that there 'is only one of me'. Ah—but now you have made a mistake—I contend that there is no rule which says that women 'love the most and miss the most'—and in one case I know that it can scarcely be true—or else you would have burst a long time ago. *I* very nearly have done, and I've calculated that another pound of love to the square inch would have done me in.

[86]

Do you ever feel full like that—have to do something to let it off—go on the ice so to speak—feel as if you would like to stiffen up and die if it would be just useful to somebody. Well men *can* feel that way and I pity those that don't.

I know dearie that you have had enough to bear and I don't wish you more.

So here is the end of your precious epistle.

8 July 3.30 am

It is night watch again and I have your last letter to answer. Well, Darling, I am glad to find you have not grown cold in my absence. I *would* like to be in your arms now—wouldn't we hug each other? So 'the others' would like to see me back also!

I am glad but also sorry to hear that you missed me so much—sorry because I feel that I have caused even more anxiety this year. If only I could chase it all away—If that I could breathe a word in your ear. You do not want me to go away again! Well I hope it will always be like that. I am sure my feelings are homing enough just now. Dearie had we been so much attached at an earlier date than Dec. 1910, for instance Dec. 1909, I would not be here now; but when I start out on a thing, it is a very satisfying feeling to my soul to do it. In this I feel something like what somebody else must have felt when they said something like this

I could not love thee so well
loved I not honour more

This Everlasting Silence

But when this undertaking is done then will be the next ones turn which was, if I remember rightly, to marry a very sweet girlie whom we both know—and that has never seemed sweeter than now. The only thing I don't like is the getting married but like everything else the 'operation' will be over in time. Would that I could do it under anaesthetic and wake up married—that might not be safe however as they might palm off somebody dreadful on to me.

It was *very* nice of you to send that box of presents—the sweets I've eaten—the mascot will be duly returned to its owner—the tie and links are really nice and are all I have in the way of civilised clothing to return in unless Eitel gets a brainwave and sends my box along. In any case I expect that what I left at Hobart will be moth eaten. Again many thanks for them. The photos are splendid—I have one of yours framed and hung over my pillow as a mascot[3]—and what do you think is on top of that again? Your chicken. So I am being well looked after. I often take 'you' down and hug you. Your Mother's photo is good also. The snaps are nice to have and have gone over them many times.

So Lica and Mary remain in Vienna—I saw in the *Bulletin* that they had gone to the seat of the war, volunteering. Shouldn't think there would be much gynaecological work doing there—However, perhaps you meant that they were to specialise in guineaology, the science of raising the guineas from patients however and by whatever means possible.

I am sorry to hear that Mary has not got that degree she went for—not because of any credit it may have been, but because she deserves a *lot*.

November 1912 to November 1913

Am glad to hear your report of Liesbeth—I saw so little of her as to form no final opinion, though naturally had a warm feeling even before meeting her. In fact you are all nice and *more* than nice. I wish you were all here now, or rather that I was in civilisation with you. It will be very interesting for me to meet Theo again—that is so long as he does not want to try hypnotising me.

So you are really more enlightened now? Well, as for me, I suppose I have gone back 2 years behind the times. 'She is wiser in many things and loves you much more than ever'. Well the last is the thing I particularly want you to get wise about—I almost feel that I could stop away longer if you would continue in that way, but I think there must be a limit—either it would run to undeserved love or, on the other hand, prove the truth of the saying that 'Distance makes the heart grow fonder of somebody else'.

You certainly have seen a number of places. Well I guess we shall see more than Paris & London together. You *have* been delving in a lot of things, n'est ce pas? That is about all the French I can remember just now.

Re your warming me every night—there is something in that. If ever I feel dull at night and can't sleep (one gets very stale cooped up here in the hut) I always think of you and a nice happy feeling comes. There is one thing the extra year down here will do, that is to relieve the rush in publication on return, for I am getting some of it done here —however, there is such a lot to do that I shall be real busy with it [until] say the end of August 1914. I feel sure I will return fatter and healthier than when I left: The trial on the sledging journey, after the accident, gave me a big set back, but it is just about all passed now—my hair has come back fairly decently also.

I say, I hope I shan't deserve the scolding and I do like looking forward to the cuddling. Yes I can stand it: I was going to say twice over but bigamy is of course not allowed.

Yes I will do my best. Won't it be grand? Can you really? How sweet of you! Yes Darling this is certainly so, and mine to you in thick red drops.

So much for those four valuable letters. How very nice it would be if only we could be together for ever so little from time to time this year. What a lot we have to talk about—I really think we shall end by not saying anything —after all, by holding my lips pressed to yours I can feel all you have to say and then the quiet (and of course with my eyes shut) allows of so much more intensiveness.

Good Night My Darling—All love and happiness which I hope to bring you
    Douglas

---

[1] SY *Nimrod* was the 136 foot barquentine in which Douglas had travelled to and from Antarctica with Shackleton's British Expedition, 1907–09.

[2] He would indeed, in 1921, be appointed Professor of Geology and Mineralogy at the University of Adelaide, a position he held until retirement in 1952.

[3] The photograph which Douglas framed, using timber from the ubiquitous packing cases, was a professional portrait taken in Holland in 1912. Showing Paquita as an elegantly dressed, poised young woman, so different from the unsophisticated girl he had left behind, it was an interesting choice to have hanging above his bunk. Now on loan from a great-grand-daughter, the

photograph can be seen today—still in that frame—in a display at the Waite Campus, University of Adelaide.

During July the weather conditions in the vicinity of the Hut were particularly severe. The velocity of the blizzards was sometimes well in excess of 100 miles an hour, with a wind average for the month of nearly 64 miles an hour. Douglas later wrote, 'Of one thing we were certain, and that was that Adelie Land was the windiest place in the world', a statement substantiated by regular observations made by the expeditioners, frequently under the duress of the very weather conditions they were recording.

But in the shelter of the Hut, and the comparative quiet of the nightwatch hours, Douglas found solace in communing with his beloved.

Winter Quarters
15 July 1913, 3 am

My Darling Paquita

I simply love these hours that are devoted to writing to you. You are divinely sent to me to make life happy, and your influence is a power even here: Can you guess how much more potent when I am at your side? When I feel your breath? When lip touches lip? How great a thing love is.

Your fresh and healthy girlhood—your trust—your love —your tenderness—All these things are ever before me, and in this frozen, austere solitude loom up as giant angels.

My Dear I only hope that I will do you credit—and would be glad to be back in civilization, for this place can only sere my body and mind making them less fit for an offering to your imperial love.

You once said 'but I can love'. How true those words have been.

Dearie my mind has been hurt again this last week. Shall I tell you what has happened? Well, it is said in a few words but so sad and has had its depressing influence upon us. It is this—Jeffryes the wireless operator left here by Capt. Davis has lost his reason—I trust it will not be permanent, but it looks bad for he is not strong brained and should never have come down here.[1] Capt. Davis should not have brought him but only good was meant. He has had very much less strain on his mind than anyone else and so we conclude must have had a weak brain originally. He has to be watched the whole time.

Do you know where I would like to be just now? In your arms, my head on your breast, all care forgotten. Oh, Dearie, is it too much to ask that Providence will some day grant this? At times it looks so far away.

Think how gloriously near each other we might be had Capt. Davis waited another 10 hours before leaving, as it was then calm and we could have gone off.[2]

How great is fate! What small things come between the highest flights of happiness and the lowest depths of despair. What small things may turn a life from a delectable path in the radiance of a summer sun to the stygian gloom of a dungeon doom. If by the Grace of God we are restored to each other, I pray that no further accident, misunderstanding, negligence or other act, word, or deed may arise to sever we two for even ever so brief a space— and when the times comes my future existence will, I feel confident, be made fuller in the bond—be it husband or brother.

I am just feeling a little bit serious tonight you see. However the time is passing and in 5 months we should be away from here—and then!

Darling kiss me as I you in token of our deep wove bond.

Douglas

[1] When Jeffryes applied in 1911 to join the Expedition, Douglas— with his usual astute assessment of character—declined to have him. Then in 1912, Jeffryes applied to Eitel, Edgeworth David and Davis without disclosing his previous unsuccessful application. In *Mawson of the Antarctic*, Paquita wrote: 'As his credentials were excellent and wireless communication with Adelie Land had until then been so unsatisfactory, [Davis] was naturally glad to engage [Jeffryes]'.

[2] Despite this expression of regret, Douglas later told the press that he doubted if, in his critically debilitated state, he could have survived the journey home in the *Aurora*, had Davis been able to wait for him. He considered that the enforced rest of 1913 had probably saved his life.

In his later publications on the AAE, Douglas refrained from criticising Davis as he did in this private letter to Paquita. In public, Douglas was loyal. In *The Home of the Blizzard*, for instance, he wrote (of the *Aurora*'s reappearance in December 1913): 'an unusual sound floated in . . . and the next moment in rushed Captain Davis, breezy, buoyant, brave and true . . . His cheery familiar voice rang through the Hut as he pushed a way into the gloom of the living room. It was an indescribable moment, this meeting after two years'. Keeping in close touch over the decades until Douglas's death, the two men remained friends, despite some serious professional differences.

Commonwealth Bay
22 July 1913, 4.30 am

Angel,

I have just had my weekly bath and feel sufficiently civilised to address myself to you. How are you to night? I do so want to know. Well, I have not not gone bug yet and our patient is somewhat improved, so things are hopeful down here.[1]

Surely nothing can come between us now, and so being I have just bethought myself that within a year we shall be something more to each other than the mere word. How curious it all will be—to think that I have to bury some 30 years of my life and start on a new line—I wonder how I shall take to it? Do you think I shall like it? Did you see me wink? My word your Douglas is going to have the time of his life—I am just looking forward to your cuddling (am not certain how to spell it). I wonder where we shall meet first. If it is on the ship I shall have to retire to the Captain's cabin and wait for your ladyship for am sure I would be much too shy to receive you adequately under the public gaze. Perhaps this is not the proper way to put it.

It is just possible we may make Adelaide.

November 1912 to November 1913

The house question has just crossed my mind—we will never find a house in Adelaide that I will like (that is, before we go into it). I am rather gadgety. I have a saving quality however that is of never being discontented. I hope you don't look for qualities in me for I really haven't got any, except perhaps obstinacy, and that is not quite a thing you would expect to write to your girl about. I would be more glad, had I not the book and lectures to think about—just abandon myself to a care-free tour with you. But then, Dearie, we will make all life a tour—from state to state of happiness. What could be more delightful. If Cook's[2] could but advertise trips like that would there not be a rush on them?

The fire is warm and I am getting sleepy. Would that I could lie down with my arms around you and thus go off to sleep.

Good Night,
Douglas

[1] Jeffryes's mental and emotional instability waxed and waned, causing Douglas and his companions a great deal of anxiety and frustration.

[2] Cook's tours founded by Thomas Cook (1808–92) in England in 1841.

S.S. Malwa[1]
17 August [1913]
To Melbourne

My more than loved Dougelly!

Liesbeth & I are travelling to Melbourne [from Adelaide] to get ready our house there for Mother & the others. You know ship life, however short—it means just longing. Liesbeth is also feeling this & as she is pouring out her feelings in a letter to some one, I cannot keep from you. Oh Dougelly, is this long long separation ever going to end! On the sea we seem a little nearer. Thought of you fills my whole time—reading is impossible. How I long to hear everything, to feel everything that you have gone through. Douglas, there is so much happiness in store for us. And if there is worry also—I shall help you ever so much better than before you left.

I'm afraid I am a very impatient person though I should be nothing but thankfulness. You will not go again, will you? I know you will not. It is not anything for married men to do. I think I am more sorry for Dr Wilson's widow than Lady Scott. But how terrible that disaster was.[2] Come back safely. Oh darling do be careful of your dear, dear self. I wonder if you have felt my love coming to you.

How happy we shall be when you return. I'm returning to Adelaide in February (if Capt. Davis still thinks it probable you shall land here or Fremantle) & shall be waiting for you. I warn you that I shall board the *Aurora* on the absolute earliest opportunity. After all these months I claim that! Shall see what I can do with Captain Creer—he generally meets incoming boats & you'll have to have a pilot. So come *home* at a nice time of day & the first or one of them to greet you will be your own me. Don't be afraid I cannot control myself! I am not the too-often tearful person you know of old! Just to see you, feel that you are really back from that perilous fascinating ice land.

Are you frozen? In heart I mean. Am I pouring out a little of what is in my heart to an iceberg? Oh, for a few private dear words. Why haven't you sent me a few coded words & trusted to my finding it out! Can a person remain in such cold and lonely regions however beautiful & still love warmly? You were not in love when with Shackleton.

How I long to hear about all this. That you love me just as much. Lean over the *Aurora*'s side and say it to the breeze—perhaps I shall hear it. Don't laugh it isn't a laughing matter. I love you to distraction & if when you return you find I am too warm! Well I can't help it. I own now I was rather cold before you left through ignorance of everything. Oh dear let's get to business!

I'm at last going to meet your Mother & Willy & Jessie in September. Am looking forward to it very much. Capt. Davis arrives by the *Orontes* next week.[3] We shall see him in Melbourne. Professor Henderson is writing to Prof Masson and Skeats, so I shall probably meet them also.[4]

I told you before I think that we were going to move & shall send you a wireless from Campbelltown. That wireless *is* a boon, although it is rather unsatisfactory. I've sent more than I meant to but did it when I was feeling very low and lonely so excuse. I quite understand your not wanting to use it for private purposes though everyone expects you to & always asks 'I suppose you hear every week!' Oh soon the need for wireless will be past. Dougelly, *Dougelly*.

With my whole heart & being I love and want you.
  Yours for *always* Paquita

This Everlasting Silence

[1] Travelling from one state to another by overseas liner was not unusual in this era. The P & O liner *Malwa* was a sister ship to the *Medina*, which in the previous year had been used as a royal yacht to take King George V and Queen Mary to India in great style.

[2] A wireless message of 22 February 1913 had informed Douglas of the death of Captain Robert Scott and his party. Douglas noted in his diary: 'I know what this means as I have been so near it myself lately'. Three weeks later, through Professor David, Scott's widow, who was then in Sydney, sent a message to Douglas: 'Love and sympathy, come back safe'. The following day, Douglas responded: 'I wish you to accept my warmest feelings of sympathy for your great loss. We in Adelie Land are full of admiration for Captain Scott's magnificent achievement. The Delprats are in Europe. Many thanks for your kind message'. Douglas later reinforced these sentiments with a letter of sympathy for Kathleen Scott, with whom he'd had contact in England while preparing for his Expedition. Yet not once does he mention the Scott disaster in his letters to Paquita. Perhaps it was there-but-for-the-grace-of-God. Or did he hope to spare Paquita the worry, wishfully thinking that she may not have heard of the death of Scott's party?

[3] Davis had been to London to solicit funding for the relief expedition—to enable him to take the *Aurora* back to Antarctica to rescue Douglas and those who stayed to support him.

[4] All three knew Douglas well, both professionally and as a friend. This gesture of Henderson's would have been to help Paquita feel part of that fraternity. And probably her knowledge of Captain Davis's movements emanated from Professor Henderson.

And so, as the southern spring set in, the lovers continued to write to each other—Paquita with a reasonable degree of confidence that her letters would reach Douglas before Christmas; Douglas only with a hope that, if the *Aurora* returned safely for them, and if he survived in the meantime, he would deliver all his letters to Paquita in person. It was a strange basis for any correspondence. For an exchange of love letters it was a supreme test.

Commonwealth Bay
15 September 1913, 8 pm

My darling,

I have not been able to settle to any writing or regular work for a fortnight—have wandered about feeling wretched. Why? Because I can't get a message through to you! It is worrying me very much. Let me tell the whole story.

Well! Since writing my last Jeffryes improved, and though not properly minded yet was comparatively right when we at length succeeded in getting the mast up again—that was a month or more ago. I had anticipated having to get Bickerton to send messages but as he is only a beginner could not expect to get much through. However, decided that Jeffryes could probably run it all right as it was a routine duty and his form of insanity [would] probably not interfere with such. He was anxious to go on and resented intrusion, so for the peace of all it was further desirous to let him resume his duties. Had he been frustrated he would likely have damaged the gear.

Everybody here clamoured to send private messages so I had to state that no private messages would be sent until a few urgent prelim. business ones were got off. In most of these I sent up 'all well' so that you would know that it was so. In one to the Registrar, Adelaide University, I stated 'All well, tell Delprats' so you would know. Then I posted a notice to the effect that each man could send one private telegram and I would send mine last (God forbid that you should think that so I valued it). I anticipated getting all through within a couple of days—It is now weeks since and my message has not gone. It was some time before we detected that Jeffryes was worse. I suspected one night that he was not playing the game, so accused him and extracted the confession that he had been trying to get through a message to the effect that 5 of us were insane— a nice message to send to the World. I do not know whether it got through or not. He sends too fast for any of us amateurs to read him. He was very contrite and took an oath to the effect that he would do his best in the future to follow orders implicity.

Since then he has done nothing but send his own telegrams. I suspect him to be playing a game, he sits at the instrument all the evening and says nothing is coming through—What can one do with a lunatic? If he really is trying to the best of his half-witted ability, to displace him may mean to get less through.

I do hope that you are not unduly anxious.

It is quite a time since I last wrote and the sun is quite high at midday but the wind, drift and low temps. (−29° today) still last. One is practically locked up the whole time. Oh for a glimpse of you and your people. I try not to be anxious—probably just as you do—but it is no use.

My own Darling there is heaps and heaps of love for you frozen up down here—Oh, for the Summer to thaw it out and give release to rush to its consignee.

Douglas

[Melbourne]
21 September [1913], Midnight

My very dear Douglas

The family is all asleep & I have a longing to say something to you. I saw the *Aurora* this afternoon. She is lying beside the dry dock awaiting her turn. Capt. Davis was very nice & we had tea in the wardroom, with Murphy & Harrison.[1] The former, driving us to Williamstown & back in his car, told me lots of little details about you all there. He told me more of the terrible time when they waited for your overdue arrival. Capt. Davis certainly had a great responsibility. Murphy says he is inclined to think too much on the dark side but I fancy Murphy is inclined to the opposite. I certainly always have a feeling that I must cheer Capt. up & say he did the best. It was a blow to have to leave you. He thinks a lot of you.

He has, with others, been very active too. £5000 from the Commonwealth & free docking. Don't worry about debt. When once you're back & have lectured, I'm sure donations will roll in & even if they don't you're better off than Shackleton, who by the way is off again South.

Douglas, do you want to go again? You don't know what it is meaning to me not to be able to hear anything from you. My pen has not its usual fluency in writing to you. It seems like writing to a wall. I want to know the trend of your thoughts & whether—oh I suppose you do though. There is no reason why you shouldn't like me as much as before. But this everlasting silence is almost unbearable. I don't want to doubt you dear but I'm afraid of the fascination of the South. All the members say they would go again & here is Shackleton off again. Will a calm life ever satisfy you? I have seen unhappiness where I thought all was well. Calm homes also have skeletons in a cupboard it seems. I want you to reassure me that all will go well with us & our love.

I long for our meeting but in a faint way dread it. Do you feel hurt at that sentence, or do you understand it? Will we feel a little strange at first I wonder?

I want you to send me one more wireless. One on receiving the post. Tell me whether you would rather I did not meet you on the wharf at Hobart. Shall I wait at our hotel? I thought at first I'd like to come on with the doctor or pilot but now I've seen the *Aurora* its better not. There is not room. Will you send me a few words so that I understand where to be. You have used the wireless so little for private use that it won't matter.

The window is open & I'm very cold. Hence the writing. I face due south out of it. I wonder is there too much ocean between us for you to get my nightly messages? I wonder why I feel so sad tonight. Am I afraid there is not room in your heart for the expedition & I?

Don't go too far west after leaving the main base. They had a hard struggle to get back last time. I can't do without you very much longer.

November 1912 to November 1913

I shall be able to contain myself at Hobart—don't be afraid of that. I am not a child any longer as before you left.

I wonder will you sledge at all before the *Aurora* arrives. Leave those instruments—they'll be covered & something might happen as you search.[2]

Its time I went into bed. It looks so cold and uninviting. My knees and arms are frozen. But I'm too cold to shut the window.

Oh Douglas don't *don't* let Antarctica freeze you. If I only had some words to go on with. Just another caress to think of. I need it so much. It would be selfish if I wished you missed me as I do you. But men don't love as women. Not as this one, any way.

Your Paquita

[PS.] I sat on your bunk & chair & red cushion this afternoon. I wish you were here to warm me. I'm in my dressing gown & my teeth is chattering. M'bourne *is* a cold place.

---

[1] The kindly friendliness of men like John Davis, Percy Correll, Charles Harrisson and Dyce Murphy was a foretaste of Paquita's lifelong contact with and interest in the former expeditioners. Yet despite her outgoing personality, did Paquita's pride prevent her from confiding to these men her deep disappointment at hearing so little from Douglas? How else did Davis not remember Douglas's letter, still lying in the *Aurora*—undelivered?

[2] In late November 1913, Douglas with two companions did try to recover the instruments left by one of the sledging parties near Mount Murchison in the previous year. Paquita's knowledge of

this intent two months before the abortive attempt, together with other references in her letters of 1913, reveal how intently she had listened to all that the returned expeditioners—and others such as Professors David, Henderson and Masson—had told her of life and problems in the South.

So much for Spring! The conditions in Melbourne seemed almost as cold as in the Antarctic—or, rather, the doubts in Paquita's mind were chilling her ardour. And little wonder. It was now nearly twenty-two months since she'd seen Douglas, and seventeen and a half months since she'd received a letter.

At the end of September, Douglas sent Paquita another wireless message:

> EVERYTHING SPLENDID HERE LOOKING FORWARD TO RETURN WITH GREAT EXPECTATIONS PLANS NEXT YEAR REMAIN SAME AS SHOULD HAVE EVENTUATED THIS YEAR LOVE DOUGLAS

While this communication should have eased some of Paquita's concerns, she was still not placated. The long silences were causing her to doubt Douglas's love for her. She was not to know how longingly and caringly he was writing, and how much he, too, was feeling 'this everlasting silence'.

Paquita considered that Douglas's 'everything splendid here'—while well-meant—was somewhat patronising. 'I'm not a child', she thought, but was too proud to confide in anyone but Douglas himself. If only she could *see* him! Or receive a letter.

She had heard through the grapevine of committee members/returned expeditioners/Captain Davis and of course Mr Eitel, that Douglas was now having extreme problems with the sadly disturbed, unbalanced and anti-social behaviour of Jeffryes. 'So why tell me "everything splendid here"?'

Williams Road
Melbourne
1 October [1913]

My darling

Your wireless came on Friday just as I returned from Prof & Mrs Skeats & all your rock specimens!

So you are looking forward with expectations to your return. So am I. May we neither be disappointed. In five months from to-day I shall be in Hobart & you should be there also. I do hope you manage to *get* the wireless apparatus from the main base on the dear old *Aurora* & that it won't be too light to send a message to your me that you're safe & *coming*. Have you any mixed feelings in the thought of seeing me again? Or is it all joy? You know my feelings for you are stronger & warmer because in my letters last year & in these I can assure you of that. But I have no such assurance. I have only the three or four wires. And oh I can't help saying it—I would have been so much happier if I had been able to think that your need to send word to me,

This Everlasting Silence

however little, more often than four times in a year, was greater than your dislike to use the wireless for private use. You can explain it all when you return I know—& I won't mind then but now with everyone asking 'do you hear every week?' its just a little worrying. I know you can't send every week that would be nonsense but none since last May until the other day. Shall I tear up this? Perhaps I shouldn't say this to you. I should wait calmly till your return & not worry you. Perhaps its only jealousy! Liesbeth got 3 big fat letters yesterday & was flushed with happiness all day. I'd rather you were away 4 years and wrote than $2\frac{1}{2}$ and no news. No darling I don't mean that. When you return it will be alright.

Let me tell you something. I didn't know anything at all about the dark side of life. I had thought such things only happened in books & not among people one knew but now I know differently. I have seen & heard more about it now and can you wonder that it has worried me? And therefore I cannot help speaking like this. I cannot take anyone but you into my confidence that life frightens me sometimes & you are the only one who can reassure me.[2] So bear with my troubles!

I think this way—that you want to know what I'm thinking & doing this year as much as I wish to know your thoughts. We must always be open & confidential. See—it has cheered me considerably writing to you! After all, only five months more. Our future *looks* happy enough. Why shouldn't it be so? What wouldn't I give for just one nice fat bulgy lovely letter! I feel quite equal to copying your handwriting & writing myself a love letter!

November 1912 to November 1913

How is your book getting on? I meant this as a surprise for you—I have taken up shorthand & typing! At least I had. I can take down letters already & type at quite a speed without looking. With a little more practice, I might get quite useful! I started in Adelaide & liked it but the college here was too dirty & common for words & I left. So now am at it alone.

[Hello my ducky Douglas! I am more clever than you!]

Now you don't know what that is *do* you? Aren't I clever?

I'm reading Amundsen.[3] Mrs Masson lent it to me. She *is* a dear. Mrs Masson & I *do* like Prof. Skeats. I saw all your geological specs. & understood quite a lot.[4]

Liesbeth gives a big concert here on the 21st. She wants to teach next year. I hope it will be a success. Leinte will be gone when you return & Lica & Mary will be en route for this paradise! I hope to go to Campbelltown this month after the concert, if Father lets me. He seems to like all his family to stop at home always!

Now I'll end. I am quite better now. That is proof I love you isn't it?

Good-bye my own. This is my fourth epistle.[5] I'll write one last one later. Five isn't too much is it? Five in one long year? You *needn't* read them you know!

    Yours with all her loving heart
    Paquita

This Everlasting Silence

1. When the Delprats moved from Adelaide to Melbourne, it was to a large home, Linden, on Williams Road. The area was sometimes referred to as Toorak, sometimes Windsor, and at other times as Prahran. Although Linden has been demolished, the site has been marked by a sub-development known as Linden Court, Prahran.
2. Was this another veiled reference to the subject of sexual relations? With none of her sisters married at this time, her two medical sisters far away in Europe, a mother who was loving but apparently reluctant to talk intimately, and instructional books not readily available to her, the long years of waiting—without a kindly confidante—could indeed have frightened Paquita regarding the physical aspect of marriage. Family lore has it, however, that Paquita was later well pleased with marital life, yet the memory of her pre-nuptial terrors remained in her mind: an unpublished novel she wrote half a century later focused on the unhappiness which sexual ignorance could cause some women.
3. *The South Pole* by Roald Amundsen, translated from the Norwegian by A. C. Chater and published in London in 1912.
4. At Tormore House in North Adelaide, the young Paquita had studied geology—a then uncommon subject for a girl.
5. It was the eighth letter since Douglas left Adelaide nearly two years before—but the fourth since the despatch of the first four letters with Davis on the *Aurora*, late in 1912.

Blizzards of a hundred miles an hour early in October (which again carried away the aerial at Commonwealth Bay) and the re-appearance of auroras rendered wireless communication spasmodic at best. But on 17 October, with spring's first penguin appearing, Douglas was able to send to Davis, c/o Eitel:

BRING ALONG EVERYBODY'S PRIVATE CLOTHING

November 1912 to November 1913

This would enable them to greet civilisation with a greater degree of cleanliness and dignity—prime requisites for professional men of the pre-World War I era. Douglas felt it particularly important for him to be well dressed for Paquita's sake, let alone for the press photographers he was sure would be awaiting the *Aurora*'s arrival. Australia, the Empire, and raising funds to cover the Expedition's cost were all important to him, but above all he longed to be with Paquita. Then doubts crowded his mind. Would Paquita even be wanting to be with him?

Winter Quarters
Adelie Land
30 October 1913

Warm Heart of Mine,

I hope that it is only I that am wracked by anxious thoughts—I trust that you are well—that you are enjoying health and life—I cannot however help feeling anxious, most anxious.

A wretched fit seizes me; sets the imagination running in a doleful tune; figures you unwell, cast down by worry—or piqued and hurt by my seeming unconcern—these wretched scraps of information that have passed to you.

Ah! Did you but know how difficult to send news. Could you be here but for one moment, how all this would be spared you.

Our future is surely but a sunlit walk under a rosy pergola, and yet these dark shaddows cast upon the present do so malign the outlook that it were well could we both sleep it out in blank oblivion. Perhaps Nature's ways are the best—the sombre present to heighten the pleasures of the future to a supreme bliss unknown were they not contrasted with the murky possibilities of existence in which we now flounder—the status quo of each unknown to other.

I have no doubt of your love you speak eloquently of it. For my part, I shall address myself in life to becoming more and more worthy of it.

I was glad to get a message off to you at last, and had hoped to get some word back in reply. Not doing I became increasingly anxious, though I console myself that there is likely something at M.Q.I. and it is merely the broken communication that prevents it coming through.

Things have been much busier of late on account of the sunlight and occasional moderately calm days. However, I found time to try my hand at designing a house. Have come to the conclusion that there is a good deal in the subject of 'house design' and find it interesting. I have a fault (?), however, of assuming it a domicile incorporating all possible suggestions of comfort (and why should one not?)—the result is unfortunately always expensive.

Dinner is just called so I shall cease for the present.

I want my P very badly—or at least I want to know that all is well with you My Darling

    Douglas

At this time, Douglas also wrote to Liesbeth (Carmen) Delprat, addressing her as 'sister' and saying how much he was looking forward to seeing her again. 'What a happy gathering it will be—you must know I am very proud of all my new sisters. I am glad you returned to Australia, for your presence must have helped pass the time for Paquita whom I fear has had a very unhappy year. I trust it will never be your fate to place such happiness in the hands of a vagrant such as I'. That letter, also, returned to Australia with Douglas.

Melbourne
Sunday, 9 November [1913]

Dearest Dougelly,

The *Aurora* leaves on Tuesday & I must add a last letter to my little pile. And I do hope it will be the last letter I'll need to write you for a very long while. This separation has been quite long enough. We shall feel almost estranged. I hope you will be in Melbourne a lot. I can come & stay in Adelaide also.

Yesterday I returned from a two weeks visit to Campbelltown. I like Willy & Dally very much & we got on together very well. The kiddies are dears & I miss them.

We sent you a joint wireless—perhaps you didn't get it? I met Eitel, Hurley and Hunter. I like Hurley. They were very good & I had a private view of the film the *Aurora* brought back in 1912 when we were away.[1]

I went of course to see Mrs Mawson. Willy will have told you all about her health. She is very comfortable where she is & well looked after. She recognised me & said how she looked forward to seeing us together. Her memory is not good & the words are mixed at times but her health otherwise seems alright.

Willy & Dally are happy. I have a respect for them both. They are living in their own house now.

Dearie I hope you & I are going to be happy. There is so much to discuss before we are married that I can't write in a letter. You understand, don't you, *quite,* when I repeat that it isn't Paquita of 1911 you're coming back to? I may have changed in ways you won't like but on the whole I don't think so. We are very different in some ways—but that shouldn't prevent our happiness. Its to be 'give and take' on both sides.

I'm longing for your return to put me at rest. It is very difficult not to think of the future. $2\frac{1}{2}$ years is a very long time out of our lives. Oh well 3 months only now.[2] Heaven give that we aren't disappopinted in each other. Our wants are different now we are both older. Oh my dear man, come back and reassure me that all will go well with us. I have lost confidence, not in you, but in the future. I want your love again. It has been hard to do without it so long.

I wonder will you return to Hobart or Adelaide. In February the former would be the pleasantest but Adelaide seems the likeliest. Capt. Davis has promised to let me know. It is no good depending on you to. If you had wanted to wire you could have. To answer mine & your brothers wire would not have upset the men down there. However everyone has their own way of thinking.

Mother & I will be waiting for you wherever you return. If we know in time, of course.

I suppose the book is nearly ready. I'm longing to read

it. Am wondering whether you will be at the University next year. Am full of wondering!!! Never mind darling whatever happens we *are* going to be happy. I'm just awfully excited to think you are so near.

The parcel we sent is uninteresting but you're a hard person to send to. This cutting is from the *Bulletin*.[3] Were you contemplating bigamy? If so, you're found out!!!

My dear, dear man. Until February—I love you & the next time I say that it will be in your arms. And will be responded to, I hope. One sided correspondence is the limit.

> Your own
> Paquita

[1] *The Mawson Australasian Antarctic Expedition 1911–1913*, from which a condensed version, *Home of the Blizzard*, was made. Both films are held by the National Film and Sound Archive, Canberra. Access videos are available on loan.

[2] Wishful thinking! It was to be another four months before the lovers were re-united.

[3] Refers to press reports of September–November 1913 of a woman who had 'a slight tram accident in Melbourne'. She was helped by a Salvation Army officer and told him that she was the wife of Dr Douglas Mawson, adding that she had come from England to meet her husband on his return from Antarctica. See also Douglas's letter of 26 December 1913.

Henrietta wrote on the following day:

> My dear Douglas, Tomorrow we are going to see the *Aurora* sail. We are all very excited . . . Paquita has kept up her spirit well, but it has been a hard time . . . I wonder where you will

land and where we shall see you first . . . it is late and we have to go early tomorrow as the *Aurora* is sailing at 10 o'clock. Take well care of yourself, with much love Your Mother-in-law to be.

The departure of the *Aurora* cheered Paquita, and reconciled her to the remaining wait. Leaving the wharf, she returned to Linden and wrote her final, final letter to her fiancé in Antarctica. Paquita addressed the letter to 'Dr Douglas Mawson, C/o Capt Davis, S.Y. Aurora, Hobart', Davis having assured her that a coastal trading vessel would be carrying mail to Hobart while the *Aurora* was taking on supplies.

But still Captain Davis failed to remember that his ship was already carrying a long-undelivered letter from Douglas to her. Or had he remembered and was by then too embarrassed to bring it out?

Melbourne
Tuesday 11 November [1913]

My own Douglas,

I cannot feel there is an opportunity to write just once more & let it pass! I've written my last letter already, but here goes for a final!

The *Aurora* went off in splendid weather this morning. We all felt it a good omen that the sun shone so brightly. I did want to go with it!

The Gov. General¹ presided at a sort of official leave taking yesterday morning. And your health was drunk & they all said nice things of you.

November 1912 to November 1913

I'm not satisfied with the mail I've sent you already. The letters are not warm enough but it is hard to write. When you come back to me I think I can show you that my love is warmer & stronger in spite of any lack of warmth I may have shown in letters.

It will be a glorious day when you return. I shall not be a bit jealous of the expedition but when everything is over —you will be happy to live quietly & not dash off again, won't you?

I have no right to say I am proud of you as I have done nothing for you but there is no one who appreciates you more than I. When you think of the hard struggle you had in London & here before leaving[2] and think that it is now nearly done—and so *very* well done well you must be glad.

I really will stop this time. If it is in my power to make you so you shall be *happy* when you return. My dear dear Dougelly.

>Your very own
>Paquita

[1] Lord Denman.
[2] Refers to the frustrations of raising funds and purchasing and borrowing equipment for the Expedition.

>*Winter Quarters*
>*Commonwealth Bay*
>*20 November 1913*

My Brave Darling,

How grand it is to be so near returning to the nearer manifestation of your love, the warmth of your bosom. Ah,

This Everlasting Silence

Dear Heart, it will be a happy day that we set each others anxieties at rest, and my Dear I trust that I shall never cause you such trouble again.

We hear, by wireless, that the ship left Hobart on Nov. 15th and expect her here on Dec. 10th or soon after.

On account of the untoward delay, I shall not enjoy, as I should otherwise, the 6 weeks exploration by ship following on her arrival. There are too many chances of accident to give a relish where my Soul is pining for you above all.

How are you? You have not come into my dreams lately to tell me—and I am so anxious.

I have been trying to think that I am back and we are fitting out a 'home'. Advertisements for furniture attract me where they used never.

Always the pros and cons float past me—should we get married soon after I return and before I go to Europe or after lectures are over and the book published and exped. debts paid? Always it seems the first is the best—though, pressed by urgent business, I will be but a poor bridegroom. Of course when 'the book' is out things will be easy—say, after Sept. 1914. Notwithstanding the early rush the latter part of our visit to Europe could be spent very *cosily* and we could go & see much.

At least I feel that if you will bear with me until the book is out, that we shall have a *very jolly* time after.

I must wait to hear from you for after all women, even if they be so young, have great wisdom in their fair heads —at least I know my Paquita has.

November 1912 to November 1913

Well that is all I have got to say—I don't care whether the Turks or the Bulgarians have got wiped out—whether Fisher or Deakin is supreme[1]—so long as my Paquita is well and happy.

Your Douglas

P.S. I have made out a fresh will a copy of which will be sent to you in case I perish in short sledging journey about to be undertaken. D.M.

[1] These references to the Balkan Wars and federal politics clearly indicate that at least some of Antarctica's isolation was being alleviated by news via the wireless.

November 1913 to February 1914

As indicated in his 20 November postscript to Paquita, Douglas set out on a short sledging journey in the third week of November, accompanied by meteorologist Cecil Madigan and cartographer Alf Hodgeman.

Extreme weather conditions, with impenetrably deep snow covering the areas near Mt Murchison, contrived to abort the mission: to retrieve scientific instruments, specimens and safety equipment cached by Madigan's Eastern Coastal Party twelve months previously. Although worthwhile scientific observations were made, to add to the wealth of data already collected by the AAE, the three-week trek was a disappointment to Douglas, who was not yet as fit as he'd thought.

On the return journey, he became deeply depressed. However, the arrival of the *Aurora* on 13 December, within hours of his return to the Hut, did wonders to restore Douglas's flagging spirits—just as the ship's departure from Melbourne had cheered Paquita a month previously.

'The moment of which we had dreamt for months had assuredly come', Douglas later wrote:

> We picked out familiar figures on the bridge and poop, and made a bonfire in a rocky crevice in their honour . . . It was

splendid to know that the world contained so many people
. . . then came the fusillade of letters, magazines and parcels.
. . . [We] sat on the warm deck and read letters and papers
in voracious haste . . . No one of us could ever erase that day
from the tablets of his memory.

But all danger was not yet past. A programme of dredging and oceanographic observations was still to be undertaken, during which six weeks of hurricanes and other capricious elements pursued the *Aurora* as she slalomed amidst pack-ice and bergs. Douglas was, as ever, vitally interested in the scientific work, but freshly anxious for the safety of his companions and for his reunion with Paquita. During that treacherous journey, the fate of the *Titanic*, which had grazed an Atlantic iceberg and sunk in April 1912, was in the background of his and his companions' minds. Douglas later gave high praise to Captain Davis for his masterly handling of the *Aurora*.

As usual, Paquita was never far from Douglas's mind. He spent a miserable Christmas Day, just as in Melbourne his beloved was also aware that this was the third successive Christmas on which they had been apart. But on Boxing Day, with a temporary lull in the weather, Douglas sought solace in writing to Paquita, answering her welcome letters delivered by Captain Davis.

November 1913 to February 1914

Antarctic Ocean
26 December 1913

True One—

What a lovely time I have had reading your letters over and over.

It has not been unalloyed pleasure, for though the joy bells have been ringing through my head in admiration of you, there has been an undercurrent of anguish—that dull pain in the heart, a constraint upon its free beating—for the anxiety I have caused you.

Believe me Paquita, I would not willingly have caused you any of it—but my Antarctic plans were laid before you and I were blended, and then under those circumstances it appeared to me right to go straight ahead.

Unexpected troubles have crept into the undertaking, with the expectation of which our private claim would have had more weight in the beginning.

My very dear Girl, my winter letters will explain the absence of more wireless news. Nodoby seems to have understood our difficulties.

Darling I thought that you had at least one letter from me by the *Aurora* of March 1913, else would I have endeavoured to send you a few more words by wireless. Before sledging I wrote several letters including one to yourself and one to my Brother. These were written with the idea of delivery in case of my non-return. Such was not indicated on them however—they were merely addressed in the ordinary way, for I thought at the time that should I not be able to reclaim the letters from the mail box in which they were deposited, it would be all U.P. with me.

Well, Capt. Davis took the box off with the letters and they were all delivered *but mine*. I now find them still in the box undelivered. I forget exactly what is in that letter, but whatever it is could not have helped being some assurance to you, notwithstanding its provision for my non-return. I am forwarding the letter to you unopened. At the present moment I intend the *Aurora* to make Adelaide. When and where I shall meet you is uncertain and so I now write in case it is possible for this to reach you before I do myself.

You must know how impatient I am to clasp you in my arms and yet we are to carry out a programme along this coast for the next month or more. The necessity of this there is no need for me to explain—we have just weathered a terrific hurricane with big odds against pulling through, but the sea has now moderated and the ship itself has suffered no serious damage, though the launch and other deck gear have gone by the board.

I am still in the dark as to how the finances are but have the assurance that things are fairly bright that way —so all my private accounts owing by the expedition will be paid.

Now Dearest all else can remain till we meet except perhaps a few references to your letters. Here goes.

'Did my love help you then' My love for you and duty to you was the real insentive which finally availed in my reaching the hut—so far it helped—I shall never regret the struggle through which it dragged me.

'So young and silly. How could you love me?' Perhaps there were a few things that I wondered at, but if they were part of a large and true heart such as I knew yours to

November 1913 to February 1914

be, I was amply satisfied. When you *love a lot,* nothing is taken amiss. Then I knew you were so young and acted according to the heart, and what more beautiful could have been? What more divine than the metamorphism of the bud to assume the radiance of the beautiful rose. Rather should I beg your forgiveness for my thoughtlessness and shortcomings.

'Wireless messages . . . unsatisfactory, want a letter.' I could never have given you my heart's feelings by wireless. Had I been really incapacitated to be your Husband, I should not have reached the Hut. My love you had to trust—*that* you need never fear for unless miracles happen to yourself. Always trust me. Will you? I believe I begin to see how you have changed—I felt that any change would be so.

Letter No 2 tells me that you are on your way to M'bourne. I hope you will not feel bad about living with me in Adelaide for a few years. We may not remain there long and the vaccations are jolly so that we can get away at the worst time.[1]

My Darling as you say you would be a great strength to me in trouble but you will be pleased to be assured that I know of no reason why our horizon should be clouded—financially or otherwise. Whilst writing this I have touched wood for one never can tell what afflictions are in store for either of us.

'You won't go again, will you?' No dearest, nothing like this will happen again—rest assured.

No! I am not frozen in heart you may be sure, this is where the warm hearts are bred.

This Everlasting Silence

Re coded words—Well! we might have thought of that before I left, but in any case it is very much more difficult to get coded words through than plain English.

Believe me Paquita, I have never at any time loved anybody as I love you. *Never* had it entered my head before I met you to wed *anybody*. This is perhaps one reason why I love you so much.

Your 3rd letter tells of a visit to the ship. You are quite right, Davis is inclined to be pessimistic. Murphy perhaps errs on the light side. Oh that I were with you to keep you warm!

'Everlasting silence' indeed it has been unbearable. I do miss you dreadfully but would not have you here for all that life holds. It is my love that wishes you not here.

How I should like to have been with you on that night of Sept. 21st.

Your 4th letter hopes we put the wireless on the *Aurora*. Well, I have never had that on the cards—it would occupy too much room and time, in return for which would be only a short service (about 200 miles only).

'Only 3 or 4 wires.' You should have had that letter and, besides the wires referred to, many assurances of our being 'all well' from various quarters intimated at the end of business telegrams. I would not have been 'all well' had my love for my Darling wavered even the least bit. Perhaps they did not tell you—Eitel and David are both very remiss. Oh; I cannot talk further about the horrid wireless —wait till we meet.

You strike the right note when you say we must always be open and confidential.

You ask how is the book getting on. The publisher has done so also. I am sorry to say there is but little done towards it. I did what I could, but most of my time during this winter was occupied in keeping myself and others sane.

Re typing. I am delighted you have learnt: you will find it useful. I like you for learning but I would not care to requisition your services in that respect. I regret I cannot read the shorthand.

The Massons, the Skeats and Henderson are all very nice. I have heard of Lisbeth's success.

Your 5th letter tells me of visit to Campbelltown. So you have met my Brother and his Wife and children—and visited poor Mother. It is a great regret to me that Mother shall not pass her old age as she gave promise of a few years ago. I believe she cannot really get better now. The whole business was so sudden.

'Dearie I hope you and I are going to be happy'. My Darling, it lies with You—*Can you be happy with me.* I have aged in appearance with this strain and may not appeal to you now. My body tissues have been strained and cannot be so good. But at heart I am just the same though perhaps more impressed with your qualities. Size me up critically, and *don't* let us get married *unless* after reflection you feel nothing but attraction and an abandonment to my desire just as I feel to yours. This is *so important* for after marriage the merest indications of splits are apt to widen to become fissures and crevasses in which all hopes of married bliss are dashed to pieces. There is the 'disillusionment' which the young and inexperienced as yourself are apt to suffer most. I therefore trust that your present love includes

This Everlasting Silence

a large 'reserve stock' to meet such exigencies. I know my own mind; I have had more experience—My 'Love Account' is invulnerable.

The joint wire from yourself and Will never came. There has been scarce a message through since early October on account of daylight interference on top of all other troubles.[2]

You say you are 'wondering shall I be at the University next year'. If you got my last wire correctly it stated 'plans same only delayed one year'. That meant that We (?) go to Europe in the Autumn to publish the book. *Now* I find David, Masson etc all expect me to be in Aust. for the Brit. [Science] Ass. meeting in Aug. or Oct. (I am not now sure which) that I do not wish to attend—as I desire to be in England all that time. My plans therefore cannot be final until I have talked things over in Australia. In the meantime I have leave from the A.U. for next year— Stillwell taking my place as locum.

I am very gratified to receive the box with presents from yourself and Will—also the very nice present from Mrs Delprat (your Mother is a Darling).

Re the cutting ex *Bulletin*, curiously enough the woman has written to me, it is a remarkable letter—you must see it—apparently the production of at least a semi-insane person. So ridiculous—but seems to have caused a stir amongst many of my friends for the cutting has been received from many quarters.

The first time I read these letters, I rushed through proud for news. The second time I don't know why but I shivered

[126]

November 1913 to February 1914

& all the blood seemed to go to my heart in one great whirling eddy. Your love made me shiver.

I must turn in now and just wait and long for our meeting. Perhaps better not on the ship—really better not until I have a wash. My domestic life has been very miserable now for 2 years and I long for a clean up—ordinary clothes and a good bed—above all a carefree sleep—not dozing with one eye open as it has been for so long. Nay much better still a sleep in your arms my Love.

Douglas

[1] Despite her first years in Australia at the much hotter Broken Hill, Paquita felt Adelaide's summers acutely.
[2] Probably another reference to the mental state of Jeffryes adding to the problems of intermittent poor wireless conditions. Others, including the engineer Bickerton and sometimes Mawson himself, had manned the wireless from time to time, but none had sufficient experience with Morse code to be as successful as Jeffryes in his better moments. (Jeffryes was certified as insane after the return of the Expedition to Australia.)

On 7 February 1914 the *Aurora* and her complement of tired men finally left polar waters for the long northward journey to Australia. Wireless receiving apparatus was set up for the last part of the voyage. Although there was not room in the cramped little ship for transmitting equipment, it was a comfort to those on board to be able to receive news from other vessels in the vicinity. Interesting snippets were pinned on a bulletin board in the ward-room.

Then, on 21 February, the four-masted barque *Archibald Russell* was sighted. For Douglas and for those who'd also spent so long in isolated Antarctica, this seemed a talisman of civilisation—almost a mirage, so like an iceberg did the *Archibald Russell*'s fully rigged sails look as the morning sun shone on her.

> *Off Kangaroo I[slan]d*
> *25 February 1914*

Precious Treasure,

Just a line in great haste between rolls. The enclosed is a letter left by me at Winter Qrs. in Nov. 1912—was taken by Capt D. to Australia and I understood delivered. He brought my box back this time and in it I found the letter.[1] Don't open it unless you like—am sure I don't remember what is in it.

With regard to my other letters, you may think them expressions of a lunatic, but remember that I have been pretty close to that state this last year and make allowances.

I only wish I could write such nice letters as you do.

Yours Douglas

[1] When I found the letter of 9 November 1912, which had travelled from Antarctica to Australia, back to Antarctica and finally to Adelaide, it was in a long, plain brown envelope, marked in the top left-hand corner 'First In Order' and addressed simply to Miss F. A. Delprat, all in Douglas's hand.

As the *Aurora* proceeded up St Vincent's Gulf in brilliant late-summer weather Paquita, who had now been twenty-two months without a letter from Douglas, was also approaching Adelaide. With what excitement yet understandable apprehension must the two lovers have faced their reunion after more than twenty-seven months apart.

Epilogue

The Reunion

'MOTHER AND I arrived by sea in Adelaide in time to meet our explorer', wrote Paquita half a century later: 'Our ship reached Outer Harbour early in the morning of 26 February 1914 and I went ashore to get a paper. In the Stop Press column was the news that the *Aurora* was approaching Adelaide'.

Paquita and Henrietta went by train to the city and from there to the South Australian Hotel on North Terrace, 'where there was a message (I don't remember how it came) that I was to wait for Douglas there'. Reporters later challenged Paquita on how she'd known that her fiancé would arrive on that day, suspecting she'd had some secret communication, 'but all I knew was that about twenty-seven months before, he had said he would be back about 26 or 27 February'.[1]

Even though Douglas's return was a year later than planned, what faith on Paquita's part to sail from Melbourne to reach Adelaide at the allotted time, and what clever timing on the part of Captain Davis and Douglas, to get the date right. Or was it that Providence, which Douglas had so often felt was caring for him in Antarctica, was still hovering?

Epilogue

The long-awaited and overdue reunion soon allayed all the lovers' doubts and fears accumulated and magnified during those long periods of the 'everlasting silence', although Paquita—like others on first seeing Douglas—was shocked at his appearance. As one journalist wrote of the returned expeditioners, 'they looked so thin, their faces were so hollow and their eyes so deep sunken',[2] but Douglas tried to make light of his residual debilitation.

Rejoicing in being able to talk and laugh together again, to hug and kiss and actually see each other, the lovers were determined that nothing further would or could stand in their way.

Growing at the back of Paquita's mind, however, was the realisation that—in common with all those whose partners are public figures—she'd always have to share her Dougelly. For the lovers to spend Paquita's week in Adelaide alone was a luxury unattainable at the time; they had to cherish each hour they could spend together, some of it beside the sea at El Rincon.

Within a few days of the *Aurora*'s arrival in Adelaide, there were two major public welcomes to Douglas, Captain Davis and the expeditioners: one at the University of Adelaide, attended by the Governor-General, Lord Denman; the other at the Town Hall, hosted by the Lord Mayor, A. A. Simpson. In happy contrast to the 1911 farewells, when she'd been persuaded to stay away, Paquita was present at these public functions. She swelled with joy and pride as a message of congratulation was read from King George V at the university's Elder Hall and, on the following day, a peel from the Town Hall bells echoed down King William Street

Epilogue

and blended with the pipe organ bellowing a celebratory theme inside the building.

The wedding date was brought forward by several months; Paquita and Henrietta returned to Melbourne to accelerate plans; Douglas reported to his colleagues and supporters in Melbourne and Sydney, and was given so many parliamentary, civic and academic receptions that he had to call 'enough'. The relentless pace and the surfeit of food were proving too much for his still-weakened constitution, yet he still gave time to be interviewed by reporters. Douglas also issued press releases both nationally and internationally; saw to the business affairs of the expedition and made plans for reducing the debts incurred by the extra year and the additional relief voyage; and, having snatched some precious time with Paquita and her family at Linden, was back in Adelaide towards the end of March for a pre-wedding presentation by his university colleagues.

There, tributes flowed not only for the explorer hero, Dr Mawson, but also for his absent bride-to-be. 'We have recognised in her', said Dr Stirling, 'fine qualities of heart and mind . . . and a very gracious and dignified presence'.[3] It was Douglas's turn to be proud of the faithful Paquita, whom he soon nicknamed Munky Punky (sometimes Monkey Puss and later Munk), enjoying her playful nature in private, but admiring her dignity in public.

On 31 March 1914, only a month after the *Aurora*'s return, Douglas and Paquita were married at Holy Trinity Church, Balaclava, in Melbourne. As she walked the long aisle to the steps of the altar, the bride's slight nervousness

Epilogue

was dispelled when Douglas, with Davis at his side, turned and flashed her his infectious grin.

The romance had come full circle from the day in 1909 when Paquita had first glimpsed that broad smile in Adelaide and had fallen in love. Carrying the bride's train at Holy Trinity was Hester Berry, the schoolfriend from Tormore who'd been beside Paquita on that day five years earlier. The protracted separation, the heartaches, were forgotten.

The First Five Years

The day after the wedding, Dr and Mrs Mawson sailed for England via Adelaide, where the Tormoreans and the university community entertained them. While the four week sea voyage in the Orient Line steamer *Orama* was a valuable time for the newlyweds, they were not always alone: Captain Davis and Dr McLean were on the same ship.

Both in transit and in England, Douglas and Archie McLean continued to work on their account of the expedition, *The Home of the Blizzard*. In London there were more receptions, both public and private; Douglas was feted and quizzed by the press, addressed the august Royal Geographical Society and, as in Adelaide on his 1909 return from the Shackleton expedition, was pursued by society matrons.

With Douglas so busy, it was often left to the faithful Davis to accompany Paquita and to tutor her on the ways of the English. Apparently neither Douglas nor Paquita now bore 'J.K.' any grudge for those undelivered letters. Paquita

Epilogue

later wrote, 'I could never have got through my honeymoon without [Davis] . . . as long as I wore a hat when everyone else did, and was there in time, everything was fine to Douglas.'[4] (Paquita's account of this period of their lives—in her *Mawson of the Antarctic*—gives a significant social profile of London immediately prior to World War I.)

They had an apartment in St James Court, Buckingham Gate, where Douglas and McLean worked on the book in the mornings, sometimes in the presence of their publisher, William Heinemann. Paquita was concerned that her husband was pushing himself too hard and, when night after night they returned late to the apartment, she'd surreptitiously beat an egg into Douglas's cup of cocoa.

A knighthood, bestowed on Douglas in June, meant that less than three months after becoming Mrs Mawson, Paquita had to adjust to another title and a further increase in social commitments. Yet the pace was sometimes relieved by quiet, private times for the newlyweds in the country. Before leaving London, they paid a short but joyous visit to the Delprat relatives in Holland, and a less happy visit of condolence to the bereaved family of Dr Xavier Mertz in Basel, Switzerland.

The Mawsons sailed from Toulon for a brief return to Australasia. As they passed through the Suez Canal and the Red Sea, the August heat was so oppressive that Paquita—who never did relish hot conditions—and Douglas—who was experiencing his first real summer for three and a half years—slept on deck. World War I was declared when the Orient liner *Orvieto* was just out of Aden.

Epilogue

During the ensuing few months, Douglas lectured in Sydney, Melbourne, Brisbane and Hobart. Paquita, in the early stages of pregnancy, travelled with her husband except for the Tasmanian leg: it was the first separation since their marriage. Writing to Paquita from Hobart, Douglas lamented, 'what a pity that we can't be always together . . . I shall be glad when you're with me again. You don't know how indispensable you are'.

They travelled together for further lecturing engagements in New Zealand, and from there sailed again for England, via Cape Horn, Montevideo and Teneriffe, with a Royal Navy escort to protect the New Zealand passenger steamer *Ruahine* from marauding enemy vessels. The Mawsons arrived safely at Plymouth in mid-December 1914. Three crowded weeks in London followed, Douglas mostly working on book production and promotion with William Heinemann, and Paquita shopping for the baby whose presence was now obvious.

Still working hard to improve the AAE's balance sheet and to promote the publication of *The Home of the Blizzard*, Douglas went on to America for a further lecture tour, the intensity of which the pregnant Paquita would have found too arduous. She therefore returned alone to Australia, via Cape Town. Separated by more and more ocean, where danger accelerated daily as Germany retaliated for being blockaded by the Royal Navy, neither of the Mawsons was ever far from the other in thoughts. Douglas wrote frequently from America as he became increasingly concerned about Paquita. On 14 February 1915 he wrote:

Epilogue

My Dear Old Monkey Puss,

Your letter from Teneriffe has arrived—it is so nice to know you have got so far all right. You must now be out of danger from marauding German vessels . . . I suppose you called at Cape Town, am anxious that you shall have done so . . . am not sure that your condition would allow you to go ashore, however, I hope dear old Punkey, that all is well with you—it certainly must be a curious sensation . . . Had you been with me you could not have rushed round quite as much as I have done.

It often means an awful scramble—after lecturing in the evening we frequently catch a train after midnight, then get out between 5 and 6 a.m.

Paquita meanwhile reached the safety and comfort of her parents' home in Melbourne. When Patricia was born at Linden in mid-April 1915 (three weeks before a German submarine sank the British passenger liner *Lusitania* with the loss of two thousand passengers and crew) she'd already travelled one and half times around the world in war-torn waters! Little wonder that Paquita had a difficult labour.

It was some weeks before Douglas was able to return to Australia to make his daughter's acquaintance. The three of them were at last able to set up their first family home, a flat in the newly built, gracious Ruthven Mansions in central Adelaide's Pulteney Street, just across from the university, where Douglas resumed his lecturing position.

The domesticity suited Paquita, who revelled and blossomed in her motherhood. Douglas was relieved to have their privacy, to re-settle to academia and to organise the extensive scientific results of the AAE for publication. But these halcyon days were short-lived.

The demands of war called to Douglas, who believed that his greatest contribution could be made in England, where he hoped his scientific knowledge rather than his notoriety as an Antarctic explorer would best be utilised. Leaving Paquita and Patricia in Melbourne, Douglas sailed off to America at the end of March 1916. Travelling from west to east by train, with further money-raising lecturing en route, he finally ventured across the Atlantic, arriving in London in mid-May.

Paquita and Douglas continued to write to each other frequently and, although the seaborne news was a month or more old by the time it was read, at least there was not a repetition of the 'everlasting silence' of the 1911–14 period. On his way across the Atlantic, Douglas wrote, 'I wish you were here, we would show off together swaggering around the pitching deck!' By September, Paquita was receiving letters and cables from Douglas in Liverpool (where he had become an embarkation officer supervising shipments of high explosives from Britain to Russia), wondering if it was at all feasible for her and Pat to join him.

Although they were missing each other acutely, the decision for Paquita to face the dangerous waters which separated them was a difficult one. But with her customary pluck she rallied to the call. Sailing first via the Pacific to California, she left Pat with Henrietta and Leinte, who had set up home there for Willie to attend medical school.

Douglas wrote at length instructing Paquita on what to do if her ship was torpedoed while crossing the Atlantic.

Epilogue

Despite a last-minute cabled warning from him to wait a little longer as there was news of enemy submarines near New York, she set off, arriving in Liverpool early in 1917. Delighted to be together again, they lived at the North Western Hotel until Douglas was transferred to London some months later.

Near Bell Weir Lock, beside the Thames at Egham-by-Runnymede, they rented May Meadow Cottage where, in a quiet, romantic environment, they were once again able to resume a domestic life.* Paquita nurtured a new life within her, and Douglas, who commuted to London by train each day, was now professionally more fulfilled. He held an investigative and advisory role, liaising between the British and Russian explosive departments, with the title of Special Intelligence Officer and the rank of Major.

In the northern autumn of 1917 they moved to a terrace house in Kensington, where in September their second daughter, Jessica, was born during an air raid. Both at Bell Weir and in London, the Mawsons had many visits from

---

* In June 1950, with her friend Doris Simpson as travelling companion, Paquita made a pilgrimage back to May Meadow. 'You can't see the river from it now', she wrote to Douglas. 'the hedge has grown so. Nice to see it again'. In 1993 my husband and I visited May Meadow cottage, to hear the sounds of the Thames over the weir and to picture Douglas and Paquita punting beneath the willows as described in *Mawson of the Antarctic*. Only the absence of their little daughter spoilt their 'wonderful summer' of 1917 together. Sadly, the modest house where the Mawsons were so happy has recently been demolished.

former members of the AAE, on leave from Commonwealth fighting forces. This established a pattern which continued for the remainder of their long life together—the AAE expeditioners and mariners becoming an extended family to both Douglas and Paquita, who followed their careers and family lives with interest.

The Brighton Years

Not until the war was over could Douglas and Paquita take Jessica to Australia to meet her sister and maternal grandparents for the first time. Douglas's mother had died, and Henrietta and Pat had returned to Melbourne from San Francisco some months before the Mawsons' homecoming in March 1919. By this time Pat had been separated from her father for three years and from her mother for more than two.

Douglas resumed his academic life in Adelaide and searched widely for a suitable family home. With something of the homing instinct of a pair of lovebirds, he and Paquita eventually decided to accept a gift from Paquita's father of part of the El Rincon land at Brighton, and to build their nest on it. While waiting for their plans to grow into a habitable home, the Mawsons rented one on the South Esplanade at Brighton, thus beginning the long final chapter of their romance—after a nomadic five years, their partnership was based at Brighton for the next thirty-nine years. By 1920 they were in their permanent home in King Street, where

Epilogue

they gardened enthusiastically, planted numerous trees, entertained widely and educated their daughters.

Sir Douglas and Lady Mawson maintained a presence as a physically striking couple, prominent in a wide range of community, university and social activities, and the recipients of a variety of awards and honours. While not always completely in tune, both were resolutely attached and loyal to each other and to their daughters and grandchildren, and shared an abiding interest and pride in their Brighton home, Jerbii, and their grazing property, Harewood, in the southern Adelaide Hills. But there continued to be many periods spent apart.

Sometimes Douglas was the wanderer—two short expeditions to Antarctica, conferences, lecture tours, government advising and lobbying, and scientific field trips; at other times it was Paquita who temporarily left the Brighton nest. She had frequent visits to Melbourne, as her parents had moved to Mandeville Crescent, Toorak, in 1920. There Paquita was cared for by Henrietta during some early illnesses, and later she reversed the role and cared for her parents as they reached old age. She also became an inveterate and intrepid traveller, making extended visits to siblings in Australia and overseas. At any one time, as many as four of them could be living in faraway, sometimes exotic, parts of the world.

Although the married life of the Mawsons failed to fit the pattern envisaged in their 1911–14 resolutions of never being parted again, their letters kept them always in touch. 'Dearest old Hubble' Paquita would write, and 'Dear Old

Epilogue

Munkey' Douglas would reply, as their correspondence travelled by rail, sea, and later by air.

While both Douglas and Paquita achieved much, their separate and joint responsibilities and commitments caused one of the disappointments of the 44-year union: after the initial 1914–19 journeys, only once, in 1926, were they able to travel overseas *together*.

Her husband's advanced views on feminism coupled with her own fierce spirit of independence had allowed Paquita much more freedom in her marriage than most of her contemporaries had. But as time went on, Douglas's deteriorating health (possibly a legacy of the near-fatal privations of 1912–13) and continuing demands upon his time, made him increasingly dependent, both in body and spirit, upon the faithful but often exhausted Paquita.

Douglas died at Brighton in 1958. Paquita lived a further sixteen years.

[1] P. Mawson, *Mawson of the Antarctic*, p. 102.
[2] From an unattributed press clipping in Paquita's scrapbook, March 1914. Within a few weeks, however, the press was reporting how fit Douglas looked.
[3] Dr Stirling's speech at the university's pre-wedding presentation to Douglas.
[4] P. Mawson, *Mawson of the Antarctic*, p. 109.

After a long and distinguished career as a scientist, Paquita and Douglas's daughter Pat (Thomas)—whose nine months in the womb had been spent mostly on ships, travelling 1½ times around the globe—died late in 1999, aged eighty-four.

As this book goes to print, the remaining direct descendants of the romance and union of Paquita Delprat and Douglas Mawson are: Gareth, Emlyn, Alun, Adriana and Amelia Thomas; Jessica, Andrew, Emma, Dimity, Angus, James and William McEwin; Paquita, Sasha, Saxon and Markham Boxton; Stella, Jessica and Friedrich Vitzthum von Eckstaedt; Alison, Andrew and Eloise Lynch.

Sources

STRUCTURED INTERVIEWS AND informal conversations with the Mawson descendants and with those who knew Paquita and Douglas Mawson, some of whom do not wish to be acknowledged individually, were invaluable in my research, as was information gleaned from the wealth of letters, scrapbooks, documents and photographs held by family members. My descriptions and annotations are in large part composites of the memories and observations of these people.

Archival Sources

The papers of Paquita and Douglas Mawson are held in a range of repositories. Douglas's are principally at the Waite Campus of the University of Adelaide. Paquita's are both there and at the Mortlock Library, State Library of South Australia; the National Library of Australia; and the Australian Manuscript Collection, State Library of Victoria.

BHP Archives, Melbourne

Company records

Charles Rasp Memorial Library, Broken Hill

Company and local records

Sources

Egham-by-Runnymede Historical Society, Surrey, England
Records of land tenure, maps and photographs

Mitchell Library, State Library of New South Wales
Australasian Antarctic Expedition Papers

Mortlock Library, State Library of South Australia
Mawson Papers, including Delprat Papers
Tormore House Papers and bound copies of *Tormorean*

National Library of Australia
Delprat-Teppema Papers

Pasminco, Broken Hill
Company records

State Library of Victoria, Australian Manuscripts Collection
J. K. Davis Papers and Journals

Waite Campus, University of Adelaide, Mawson Collection
Mawson Papers

Books and Articles

Angove, D., *Tormore: A Tribute to Caroline Jacob*, Tormore Old Scholars' Association, Adelaide, 1962.

Ayres, P., *Mawson: A Life*, Melbourne University Press, Melbourne, 1999.

Sources

Barrier Daily Truth (Broken Hill), May–August 1909.

Bickel, L., *This Accursed Land*, Macmillan, Melbourne, 1977.

Boston, P., *Home and Away with Douglas Mawson*, privately published, Carnarvon WA, 1988.

Bowden, T., *The Silence Calling*, Allen & Unwin, Sydney, 1997.

Chatteron, E. K., *Steamships and Their Story*, Cassell, London, 1910.

Cleland, J. and Southcott, R. V., 'Hypervitaminosis A in the Antarctic in the Australasian Antarctic Expedition of 1911–1914: a possible explanation of the illnesses of Mertz and Mawson', *Medical Journal of Australia*, part 1, 1969, pp. 1337–42.

Corson, F. R., *The Atlantic Ferry in the Twentieth Century*, Sampson Low, Marston, London, 1932.

Crossley, L. (ed.), *Trial By Ice: The Antarctic Journals of John King Davis*, Bluntisham Books and Erskine Press, Huntingdon and Norfolk, 1997.

The Hands of a Woman: South Australian Medical Women's Society, Wakefield Press, Adelaide, 1994.

Huntford, R., *Shackleton*, Hodder and Stoughton, London, 1985.

Innes, M. and Duff, H., *Mawson's Papers: A Guide*, Mawson Institute for Antarctic Research, Adelaide, 1990.

Kearns, R., *Broken Hill: A Pictorial History*, Investigator Press, Adelaide, 1982.

—— *Broken Hill 1894–1914*, Broken Hill Historical Society, Broken Hill, 1974.

Mackinnon, A., *Love and Freedom*, Cambridge University Press, Cambridge, 1997.

Sources

Mawson, D., *The Home of the Blizzard: The Story of the Australasian Antarctic Expedition 1911–1914*, Wakefield Press, Adelaide, 1996 (1st edn, London, 1915).

—— *Mawson's Antarctic Diaries*, edited by F. Jacka and E. Jacka, Allen & Unwin, Sydney, 1988.

Mawson, P., *Mawson of the Antarctic*, Longmans, Green and Co., London, 1964.

—— *A Vision of Steel: The Life of G. D. Delprat*, F. W. Cheshire, Melbourne, 1958.

Moyes, J. L., *Exploring the Antarctic with Mawson and the Men of the 1911–1914 Expedition*, privately published, West Gosford NSW, 1997.

'Ockham's Razor', ABC Radio, June 1997 (transcript).

Oliver, H. *Mawson's Antarctic Diaries*, University Radio 5UV cassette series, Adelaide, 1990.

Parer, D. and Parer-Cook, E., *Douglas Mawson: The Survivor*, Alella Books, Morwell Vic., 1983.

Rhodes Shipping Annual and Directory of Passenger Steamers, George Philip and Son, London, 1914.

Roberts, P., *Emily's Journal: The Welch Letters*, privately published, Adelaide, 1986.

Robinson, N., *Reluctant Harbour*, Nadjuri Australia, Jamestown SA, 1976.

Thurston, E. Temple (ed.), *Patchwork Papers*, Hodder and Stoughton, London, 1910.

Watson, M., *The Spy Who Loved Children: The Enigma of Herbert Dyce Murphy 1879–1971*, Melbourne University Press, Melbourne, 1997.

Young, L., *A Great Task of Happiness: The Life of Kathleen Scott*, Macmillan, London, 1995.

Biographical Index

Bickerton, Frank (b. 1889) Oxford-born, a motor engineer and a valued general all-rounder, he was injured as an airman in World War I.

Correll, Percy (1892–1974) Having joined the Expedition as a 19-year-old science undergraduate from Adelaide, his skill with the then experimental medium of colour photography is evident within the Mawson Collection of the University of Adelaide.

David, T. W. Edgeworth (1858–1934) Professor of Geology, University of Sydney 1891–1924, member of Shackleton's British Antarctic Expedition 1907–1909, and an administrator of Mawson's AAE. He saw active service in France during World War I, and remained a mentor, friend and prolific correspondent of Douglas Mawson until his death.

Davis, J. K. (1884–1967) First Mate and later Master of Shackleton's *Nimrod*, Mawson, in 1911, appointed him second-in-command of the Expedition as well as Master of *Aurora*.

Delprat, Elizabeth Francisca Carmen (Liesbeth or Carmen) (1884–1959) Paquita's sister. A concert violinist of some note, she gave fifteen performances in Amsterdam and Berlin in 1912 before returning to Australia in 1913. She later married a Dutch diplomat, Petrus Teppema.

Delprat, Elizabeth Theodora Johanna Stoffelina (Lica) (1882–1963) Paquita's eldest sister. A graduate in medicine from the University of Adelaide, she practised in South Australia after her marriage to Dr Milo Sprod.

Delprat, Guillaume Daniel (GDD) (1857–1937) Paquita's father. General Manager of BHP 1899–1921.

Delprat, Guillaume Daniel (Willy or Dan) (1896–1965) Paquita's youngest sibling. A surgeon, he trained, married and practised in California. He was particularly supportive of Paquita after Douglas's death.

Delprat, Henrietta Maria Wilhelmina Sophia, née Jas (1857–1937) Paquita's mother.

Biographical Index

Delprat, Madeleine Prouisette (Leinte) (1889–1967) Paquita's youngest sister. She married Fer MacDonald and lived in California for the last fifty years of her life.

Delprat, Maria Johanna Alberta (Mary) (1886–1970) Paquita's sister. A graduate in medicine from the University of Adelaide, she practised in Melbourne before marrying Peter van Buttingha Wichers, Consul for the Netherlands in Melbourne.

Delprat, Theodore Daniel (Theo) (1881–1965) Paquita's brother. After working as a metallurgist, he later took up medicine and practised for many years at Mendouran, New South Wales.

Harrisson, Charles T. (1868–1914) An Expedition biologist, he returned to Australia on the *Aurora* in late summer 1913.

Henderson, George (1870–1944) Professor of History and English at the University of Adelaide and a member of the Expedition planning committee.

Hodgeman, Arthur (b. 1885) Cartographer and artist for the Expedition, with leave of absence from his Adelaide positions as architect and government draughtsman.

Hunter, Dr John (1888–1964) Biologist for the Expedition during its first year, he returned with Davis from the first relief voyage of 1912–13.

Hurley, Frank (1885–1962) Official photographer during the Expedition's first year, he returned with Davis from the first relief voyage of 1912–13.

Jeffryes, Sidney The relief wireless operator left by Davis at Commonwealth Bay.

McLean, Archie (1885–1922) The Expedition's Chief Medical Officer and Bacteriologist, he was gassed during World War I while serving in the Medical Corps.

Madigan, Cecil (1889–1947) Chosen by Davis to lead the relief party, he served in World War I, enlisting from Oxford to defer his Rhodes Scholarship for the second time. He was later a senior lecturer in geology at the University of Adelaide, once again working with Mawson.

Masson, Orme (1858–1937) Professor of Chemistry, University of Melbourne, and a member of the Expedition planning committee.

Mawson, Jessie (Dally) Married to Douglas's brother.

Mawson, Margaret (née Moore) (1859–1917) Douglas's mother.

Mawson, Robert (1854–1912) Douglas's father.

Mawson, William (1880–1939) Douglas's brother. A medical practitioner at Campbelltown, near Sydney, where his parents also lived.

Mertz, Xavier (1882–1912) A Swiss ski champion and mountaineer, with a doctorate in law.

Biographical Index

Murphy, Herbert Dyce (1879–1971) An adventurer, in charge of stores at Commonwealth Bay, he returned to Australia on the *Aurora* in late summer 1913. A book about his enigmatic life, *The Spy Who Loved Children* by Moira Watson, was published in 1997.

Ninnis, Belgrave (1887–1912) A lieutenant with the British Royal Fusiliers and the son of a former Arctic expeditioner.

Scott, Kathleen (1878–1947) Widow of Robert Scott, she shared a warm, platonic relationship with Mawson, as with many other interesting men of her generation.

Scott, Robert Falcon (1868–1912) British Antarctic explorer who, with four companions, perished near the South Pole while Mawson was on his Far Eastern sledging journey.

Shackleton, Ernest (1874–1922) Irish-born ship's officer, explorer and adventurer. He made several expeditions to Antarctica, the first under Captain Robert Scott in 1901–03 and the second as leader of the British Antarctic Expedition 1907–09, of which Mawson was a member.

Skeats, E. W. (1875–1953) Professor of Geology at the University of Melbourne and a member of the Expedition planning committee.

Stillwell, Frank (1888–1963) Geologist for the Expedition, he later taught at the University of Melbourne and was involved in mineralogy at the Commonwealth Scientific and Industrial Research Organisation.

Stirling, Edward Surgeon and Professor of Physiology at the University of Adelaide (1900–1919) and Director of the South Australian Museum (1895–1915).

Way, Sir Samuel (1836–1916) Then Chief Justice of South Australia, Chancellor of the University of Adelaide and Lieutenant-Governor.

Wild, Frank (1873–1939) A seasoned Antarctic explorer chosen by Mawson as Leader and Sledge-master of the Western Base (Queen Mary Land).

Wilkes, Charles (1798–1877) An American naval officer who, during a circumnavigation of the world, spent forty-two days in Antarctic waters in 1840.

General Index

Adelaide, 5, 6, 10, 11, 13, 15, 16, 34, 37, 59–61, 73, 82, 94–6, 107, 108, 111, 112, 122, 123, 127, 128, 131–4, 137; North Adelaide, 6, 30, 53, 57

Adelaide, University of, 3, 8, 9, 12, 48, 75–7, 85, 90, 100, 126, 132, 137, 142; Waite Campus, 69, 91

Adelie Land, 25, 26, 28, 29, 43, 46, 54, 93, 98, 109

America, 74, 76, 136, 138–40

Amundsen, Roald, 37, 107, 108

Archibald Russell, 127, 128

Aurora, SY, 15, 19–22, 24–7, 33, 38, 41, 43, 44, 46, 48, 51, 60, 64, 74, 78, 93, 96–8, 101–3, 105, 108, 109, 111, 113, 114, 119–22, 124, 127, 128, 131–3

Bell Weir (Surrey), 139

Berry, Hester, 1, 134

BHP (Broken Hill Proprietary Co.), 2, 5, 6

Bickel, L., 45, 46, 55

Bickerton, F., 99, 127

Boyd family, 2, 16

Brighton (SA), 6, 13, 30, 140, 141

Brisbane, 13, 136

Broken Hill, 2, 5, 6, 13, 16, 30, 127

Campbelltown (NSW), 42, 56, 57, 97, 107, 111, 125

Commonwealth Bay (Main Base, Winter Quarters), 44–6, 49, 50, 52, 57, 64, 79, 94, 99, 115

Correll, P., 58–60, 103

David, T. W. Edgeworth, 3, 19, 63, 71, 93, 104, 124, 126

Davis, J. K., 15, 21, 45, 48, 51, 52, 57, 59, 71, 79, 92, 93, 96–8, 191, 103, 105, 108, 112, 114, 120, 122, 124, 128, 131, 132, 134

Delprat, Paquita: birth, 4; arrival in Australia, 5; education, 5, 6, 14; meets Douglas, 2; engagement, 12; reunion with Douglas, 131, 132; marriage, 133; birth of daughters, 137, 140; death, 142; descendants, 143

Delprat family: Dan (or Willy), 6, 30, 38, 40, 86, 107, 138; GDD, 2, 4–11, 13, 16, 22, 53, 140, 141; Henrietta, 4–6, 19, 30, 31, 37–40, 43, 56, 59, 60, 80, 88, 113, 126, 131, 133, 138, 140, 141; Leinte, 2, 3, 22, 30, 32, 37, 38, 40, 86, 107, 138;

[152]

General Index

Lica, 2, 22, 30–2, 34, 38, 40, 43, 60, 80, 86, 88, 107; Liesbeth (or Carmen), 2, 30, 38, 40, 60, 89, 95, 106, 107, 110; Mary, 2, 30, 38, 40, 43, 60, 80, 88, 107; Theo, 40, 42, 60
Denman, Lord, 114, 115, 132

Eitel, C., 25, 31, 33, 44, 53, 63, 74, 78, 88, 93, 104, 111, 124
El Rincon, 6, 7, 46, 140
England, 3, 4, 8, 13, 28, 74–6, 85, 95, 98, 113, 126, 136, 138
Europe, 25, 27, 30, 34, 37, 38, 40, 43, 44, 58, 59, 62, 64, 74, 76, 83, 98, 116, 126
expedition vessels *see* Aurora, Nimrod, Toroa

faith/Providence, 25, 27, 56, 66, 92, 131
feminist issues, 62, 64, 66, 67, 142
finances, personal, 6, 8–10, 13, 48, 74–8, 85, 122; *see also* fundraising (expedition)
food, 45, 49, 50, 51, 60, 80
Fremantle, 31, 32, 34, 96
fundraising (expedition), 13, 20, 73–5, 98, 101, 109, 115, 133, 136, 138

geographical societies, 76, 78, 134
geology, 3, 85, 90, 108

Hague, The, 32–4, 36, 39
Hannam, W., 44
Harrison, C. T., 101
health, 50, 54, 56, 79, 111, 119, 125, 132, 133, 141, 142
Heinemann, William, 135, 136
Henderson, G., 59, 61, 97, 98, 104, 125
Hobart, 19, 20, 23, 25, 33, 34, 44, 45, 64, 88, 102, 103, 105, 112, 114, 116, 136
Hodgeman, A., 119
Holland, 4, 23, 30, 34–6, 38, 82, 84, 85, 90, 135
Hunter, J., 111
Hurley, F., 111
Hut *see* Winter Quarters

isolation, 5, 14, 117

Jeffryes, S., 52, 92, 93, 95, 99, 100, 127
Joyce, E., 20, 21

Largs Bay, 5
Linden, 108, 114, 133, 137
London, 4, 27, 39–41, 89, 98, 115, 134–6, 138, 139

McLean, A., 52, 54, 69, 134, 135
Macquarie Island, 20–4, 26, 31, 44, 45, 53, 64, 83, 110
Madigan, C., 52, 119
Masson, O., 97, 104, 107, 125, 126
Mawson, Douglas: birth, 3; arrival in Australia, 3; education, 3; meets Paquita, 2; engagement, 12; leaves for Antarctic, 15–20; death of Ninnis,

[153]

General Index

Mertz, 49, 50; reunion with Paquita, 131, 132; marriage, 133; knighthood, 135; death, 142; descendants, 143

Mawson family: Dally (Jessie), 60, 61, 97, 111, 112; Jessica, 139, 140; Margaret, 3, 7, 8, 12, 42, 48, 56, 71, 97, 111, 140; Patricia, 137, 138, 139, 140, 142; Robert, 3, 8, 42, 48, 56, 57, 71; William, 3, 8, 42, 71, 97, 111, 112, 121, 126

May Meadow Cottage, 139

Melbourne, 12, 13, 61, 64, 96, 97, 101, 103–5, 108, 111, 113, 114, 119, 120, 123, 131, 133, 136, 137, 140, 141

Mertz, X., 47, 49, 50, 58, 135

Murphy, H. D., 26, 101, 103

New Zealand, 3, 74, 76, 136

Nimrod, SY, 3, 81, 90

Ninnis, B., 47, 49, 51, 58

Paquita bags, 14, 50

Patchwork Papers, 23, 63, 64, 83

Perth, 13, 61

Port Pirie, 9, 11, 13

royalty: British, 19, 98, 132; Dutch, 37, 38, 86

Scott, Kathleen, 96, 98, 149

Scott, Robert, 21, 37, 52, 59, 96, 98

Shackleton, E., 1, 3, 4, 9, 19, 21, 42, 90, 102, 134

shipping, passenger, 5, 13, 30, 41, 61, 95–8, 131, 134–6; see also expedition vessels

Simpson, A. A., 132

Simpson, Doris, 139

Skeats, E. W., 97, 105, 107, 125

Spain, 5, 32, 34

Stillwell, F., 126

Stirling, E., 54, 133, 142

Sydney, 3, 12, 13, 24, 76, 98, 136

Tormore House, 2, 62, 108, 134

Toroa, 23, 24

war: Balkans, 42, 88, 117; World War I, 135–9

Way, S., 12

Wild, F., 26, 44, 45, 52

Wilkes, C., 26, 28

winds, 27, 44, 45, 47, 91, 108, 120, 122

Winter Quarters, 46, 51, 55; see also Commonwealth Bay

wireless telegraphy, 14, 31, 33, 34, 37, 43, 44, 46, 51–6, 58, 60, 63, 70, 71, 92, 93, 97, 99, 100, 102, 104–6, 108, 110, 112, 116, 117, 121, 123, 124, 126, 127

Also published by
Melbourne University Press

Mawson: A Life
by Philip Ayres

In the heroic age of polar exploration, Sir Douglas Mawson stands in the first rank. His Antarctic expeditions of 1911–14 and 1929–31 resulted in Australia's claiming 40 per cent of the sixth continent. The sole survivor of an epic 300-mile trek, Mawson was also a scientist of national stature. His image on banknotes and stamps reflects enduring public esteem.

Yet until now there has been no comprehensive, objective biography of this tall, quiet figure. Aside from his two great expeditions, we have known remarkably little about him.

Philip Ayres has uncovered a complex and interesting figure. He portrays Mawson the geo-politician with influential friends and rivals. Ayres also shows us the devoted husband of Paquita; the social Mawson of the Adelaide Club; the scientist within his national and international networks; the geologist who in 1924 failed to get the Sydney Chair; and the litigious Mawson, suing or threatening suit against associates who failed him.

The icon both converges and conflicts with the real man. In this long-awaited, most impressive and readable biography, Philip Ayres not only illuminates Douglas Mawson's many achievements but also enables us to know and understand him as a human being.

The book's many illustrations include reproductions of exquisite early colour photographs from the Antarctic expedition of 1911–14.

0 522 84811 7 Hardcover 342 pp 20 colour plates
Available at all good bookshops

in which we now flounder — the status quo of each unknown to other. If I have no doubt — I you were from Speak eloquently of it — for my part, I shall address myself in life to becoming more and more worthy of it.

I am glad to get a message If to you at last — and had hoped to get some word back in reply — but doing